Minerals, Rocks and Inorganic Materials

Monograph Series of Theoretical and Experimental Studies

8

Edited by

W. von Engelhardt, Tübingen · T. Hahn, Aachen
R. Roy, University Park, Pa. · P. J. Wyllie, Chicago, Ill.

S. K. Saxena

Thermodynamics of Rock-Forming Crystalline Solutions

With 67 Figures

Springer-Verlag New York · Heidelberg · Berlin 1973

Dr. *Surendra K. Saxena*

Department of Geology, University of Delhi
Delhi 7/India

and

Department of Geology, Brooklyn College
City University of New York
Brooklyn, New York/USA

ISBN 0-387-06175-4 Springer-Verlag New York · Heidelberg · Berlin
ISBN 3-540-06175-4 Springer-Verlag Berlin · Heidelberg · New York

To Marie Louise and Hans Ramberg

Contents

Symbols and Notations

T	temperature
R	gas constant

Symbols not included above are of local use and have been defined at appropriate places.

Components: Fo, forsterite (Mg_2SiO_4); Fa, fayalite (Fe_2SiO_4); Alm, almandine ($Fe_3Al_2Si_3O_{12}$); Gr, grossularite ($Ca_3Al_2Si_3O_{12}$); Pyr, pyrope ($Mg_3Al_2Si_3O_{12}$); Sp, spessertine ($Mn_3Al_2Si_3O_{12}$); Fs, ferrosilite ($FeSiO_3$); En, enstatite ($MgSiO_3$); Ab, albite ($NaAlSi_3O_8$); Or, orthoclase ($KAlSi_3O_8$); Di, diopside ($CaMgSi_2O_6$); Hed, hedenbergite ($CaFeSi_2O_6$).

Solutions: ol, olivine; gar, garnet; opx, orthopyroxene; cpx, clinopyroxene; pl, plagioclase; or, alkali feldspar; bi, biotite.

Introduction and Acknowledgements

This monograph is written for advanced students and research workers in the fields of mineralogy, petrology and physical geochemistry. It introduces the reader to the popular solution theories and the special problems connected with the treatment of heterogeneous phase equilibria involving complex crystalline solutions. The results and discussions in the recent publications of J. B. THOMPSON, D. WALDBAUM, R. F. MUELLER, R. KRETZ, E. J. GREEN, S. GHOSE and from my work during the last eight years, have been used extensively in the book.

Most rock-forming minerals are crystalline solutions of two or more components. In petrology RAMBERG and DEVORE (1951) first used the condition that the distribution of a component in coexisting crystalline solutions in chemical equilibrium is governed by thermodynamic laws. The significance of their theoretical discussion was not realized until KRETZ (1959) and MUELLER (1960) demonstrated the orderly distribution of elements in coexisting minerals in rocks of high amphibolite and granulite facies. Their work not only showed that a high degree of chemical equilibrium may be attained in metamorphic reorganization of matter but also that to understand the chemical processes in nature, it is important to understand the nature of the crystalline solutions.

Experimental petrologists have been aware all along of the problem of the phases of variable composition in the synthetic assemblages. If the phase diagrams are to be more informative on the pressure-temperature behaviour of natural assemblages, it should be possible to vary the composition of the phases in the synthetic assemblages. As there may be experimental difficulties, it would be desirable to assess theoretically the effect of changing the compositions of the phases on the pressure-temperature stability of the assemblage. This cannot be done unless a definite attempt is made by mineralogists and petrologists to understand the thermodynamic theory of solutions and the problems encountered in applying such knowledge to rock-forming crystalline solutions. It is expected that the material presented in this work will be useful in recognizing the mineral-chemical problems and in the development of the theory of crystalline solutions.

GHOSE (1961, 1965) made the interesting study of $Mg^{2+}-Fe^{2+}$ order-disorder in cummingtonite and orthopyroxene by using X-ray methods. MUELLER (1962) considered the $Mg^{2+}-Fe^{2+}$ order-disorder in silicates and suggested that homogeneous equilibria may be discussed using the methods of GIBBS for heterogeneous equilibria. The experimental work of GHOSE and the theoretical work of MUELLER have attracted the attention of many mineralogists lately and important papers by THOMPSON (1969), MATSUI and BANNO (1965), MUELLER (1967, 1969), VIRGO and HAFNER (1969), and SAXENA and GHOSE (1971) among others, have appeared exploring and developing the field of homogeneous or intra-crystalline ion-exchange equilibria. In dealing with the theory of intra-crystalline distribution of ions, in Chapters 7 and 8, I have expressed my own views freely. It is expected that through these chapters the reader will be able to get an idea of the problems encountered in this little investigated field of mineralogy.

Most of my work referred to in this monograph was done at the Institute of Mineralogy, Uppsala, Sweden, and at the Laboratory for Space Physics, Goddard Space Flight Center, Greenbelt, USA. I had very useful discussions on problems in chemical thermodynamics and mineral-chemistry with HANS RAMBERG, ROBERT F. MUELLER, RALPH KRETZ and SUBRATA GHOSE.

I especially appreciate the help of RALPH KRETZ, RICHARD WARNER, HOWARD DAY, ROBERT F. MUELLER and SUBRATA GHOSE for reviewing and discussing various parts of the manuscript. All the computer programs presented in the Appendix have been written by PATRICIA COMELLA at Goddard and FRANK TURPIN at Virginia Polytechnic.

The manuscript was partly prepared at Goddard Space Flight Center and partly at the Department of Geology, Virginia Polytechnic Institute. I am grateful for the facilities provided by LOUIS S. WALTER, PAUL H. RIBBE and G. V. GIBBS at these places. RAMONDA HAYCOCKS and MONA SAXENA typed the manuscript and SHARON CHIANG drafted some of the figures.

I. Thermodynamic Relations in Crystalline Solutions

Thermodynamic relations between the concentration of a component in a solution and its chemical potential and other thermodynamic functions of mixing are presented here. The details of the simplifying assumptions and the methods of statistical thermodynamics have been given by DENBIGH (1965), GUGGENHEIM (1952, 1967), and PRIGOGINE and DEFAY (1954), among others. Recently THOMPSON (1967) also considered the properties of simple solutions. Besides a summary of thermodynamic relations in binary, ternary, and multicomponent solutions, the difficulties encountered in their application to silicate minerals will be considered. Some of these problems, such as the choice of a component and definition of its chemical potential in a silicate, have been discussed by RAMBERG (1952a, 1963), KRETZ (1961), and THOMPSON (1969).

1. Crystalline Solutions

The crystalline solutions considered here are rock-forming silicates forming isomorphous series with one another. Such crystalline solutions have a definite structural framework with generally two or more kinds of nonequivalent structural sites. The type of sites and the ions that occupy them vary in different crystalline solutions. The overall crystal symmetry of a solution does not change as a function of the composition, though certain microscopic details within the crystal, i.e., the form and size of the individual structural sites, may change with changing composition.

Orthopyroxene (opx) $(Mg, Fe)_2 Si_2 O_6$ may be considered as an example. In the crystal structure there are single silicate chains parallel to the c-axis held together by the octahedrally coordinated Mg^{2+} and Fe^{2+}. There are two kinds of structurally nonequivalent sites M 1 and M 2 occupied by Mg^{2+} and Fe^{2+}. The M 1 octahedral space is nearly regular polyhedral, but the M 2 space is quite distorted. If the content of Mg^{2+} and Fe^{2+} in the crystal varies, the general symmetry of the crystal does not change, but there are distinct changes in M 1 and M 2 polyhedra. The former becomes more regular and the latter more distorted with increasing Fe/Mg ratio. Such microscopic changes at the structural sites within the same crystal framework may be regarded as continuous; and the resulting energy changes, a consequence of the mixing or solution of the species to form a crystalline solution.

When the crystal structures of the end members forming a crystalline solution are the same, the cations occupy similarly coordinated sites and the cell dimensions are of comparable magnitude, we have a complete crystalline solution series over a wide range of temperature. There may also be crystalline solution between structurally different crystals. In such cases, there is a critical temperature below which the solution may usually unmix into two different crystalline solutions, each with a different crystal structure. Plagioclase ($NaAlSi_3O_8$–$CaAl_2Si_2O_8$) may be considered as a crystalline solution of this type. At higher temperatures (700° C and above) the two end members form a continuous solution series with the structure changing continuously from a Al–Si disordered end member (albite) to the Al–Si ordered end member (anorthite). At lower temperatures ($\sim 600°$ C and below) there is a complex unmixing into intermediate crystalline solutions.

2. Choice of a Chemical Component

The definition of a component in a mineral is not unique. The components in orthopyroxene may be considered to be the molecules $MgSiO_3$ and $FeSiO_3$ or MgO, FeO, and SiO_2 or the ions Mg^{2+}, Fe^{2+}, Si^{4+}, and O^{2-}. In petrological studies, the choice of a component is determined by known or postulated chemical reactions involving a mineral. In such studies, the use of components such as $FeSiO_3$ or FeO is convenient, even though there are no discrete units of this kind in the orthopyroxene crystal structure. However, when the thermodynamic properties of silicate crystalline solutions are being considered, it is only realistic to consider the ions as the components (see BRADLEY, 1962). Indeed, it can be noted that if the substitution of the cation Fe^{2+} by Mg^{2+} in orthopyroxene does not produce any changes in the silicate framework, or if there are any slight changes, they are directly a function of the changing Fe/Mg ratio, the alternative methods of defining $FeSiO_3$ or Fe^{2+} as a component are equivalent (see also SAXENA and GHOSE, 1971). The same, however, is not true in some other crystalline solutions such as plagioclase feldspars, where the change in Na/Ca leads to both structural and compositional changes in the Al–Si framework.

3. General Properties of Solutions

a) Chemical Potential

While describing the mixing properties of solutions, we shall be very much concerned with the partial molar Gibbs free energy G_1 of a component n_1 given by

$$G_1 = \left[\frac{\partial G}{\partial n_1} \right] T, P, n_2, n_3, \ldots . \tag{I.1}$$

G is the total free energy of the system. G_1 is commonly abbreviated by μ_1 and is called the chemical potential of component n_1. The chemical potential μ_1 is clearly the total response of a finite system to the addition of an infinitesimal amount of constituent n_1 when the process is carried out at constant T, P, n_2, n_3, etc. μ is an intensive variable and depends only on P and T and is independent of n. μ in terms of work may be recognized as the increased capacity of the phase per mole for doing work because of the addition of an infinitesimal amount of the constituent. We may express the fundamental equation which relates internal energy U, entropy S and volume V as

$$dU = T\,dS - P\,dV + \sum \mu_i\,dn_i \tag{I.2}$$

μ_i may also be expressed as

$$dG = -S\,dT + V\,dP + \sum \mu_i\,dn_i \tag{I.3}$$

$$dH = T\,dS + V\,dP + \sum \mu_i\,dn_i \tag{I.4}$$

Chemical potential of a component in solution is a useful quantity because the chemical potential behaves as though it represents the partial molar Gibbs free energy of the component as it exists in solution. In other words the chemical potential of a component may be regarded as the part of the total Gibbs free energy of the solution attributable to the presence of this component.

b) Ideal Crystalline Solution

The ideal solution model is a very useful reference state, particularly for the crystalline solutions. Here, unlike liquid solutions, we can conceive of the mixing of two or more atoms A and B by counting the number of sites that may be occupied by atoms of A and atoms of B in the mixed crystal. Let us consider that one pure crystal consists of N_a atoms of A and the other pure crystal N_b of B. In the mixed crystal lattice both A and B will be present and if the two atoms are closely similar in regard to their volume and forces and the two crystalline frameworks similar in geometry, the general geometry of the crystal framework will remain essentially unchanged and in the mixing process we have

$$\Delta U = \Delta H = 0 \tag{I.5}$$

In order that (I.5) may be valid, it is necessary (a) that the original crystals and the solutions have the same structure, and (b) that the atoms are interchangeable between the crystallographic sites without causing any change in the atomic or molecular energy states or the total volume of the system. With these assumptions, the entropy of mixing will be given by (see DENBIGH, 1965)

$$\varDelta S = - R(x_a \ln x_a + x_b \ln x_b) \tag{I.6}$$

where x_a is the atomic ratio $N_a/(N_a + N_b)$ and similarly x_b. The above expression for $\varDelta S$ results from the expectation that if (a) and (b) are true, the distribution of A and B will be random.

The change in Gibbs free energy is

$$\varDelta G = \varDelta H - T \varDelta S$$

By using (I.5) and (I.6) we have

$$\varDelta G = RT(x_a \ln x_a + x_b \ln x_b) \tag{I.7}$$

Let μ_a and μ_a^0 be the chemical potentials of A in the mixture and in the pure crystal respectively both at the same P and T. Then from (I.1) we have

$$\mu_a - \mu_a^0 = \left(\frac{\partial \varDelta G}{\partial x_a} \right) T, P, x_b \tag{I.8}$$

or

$$\mu_a - \mu_a^0 = RT \ln x_a$$

$$\mu_a = \mu_a^0 + RT \ln x_a \tag{I.9}$$

which may be used to define an ideal solution. A solution is ideal if the chemical potential of every component is a linear function of the logarithm of its mole fraction according to the relation (I.9).

Gibbs free energy G of a solution may be represented by

$$G = \sum_i (x_i G_i) + G^M \tag{I.10}$$

where x is the mole fraction, G_i the free energy of a pure component i and G^M the change in free energy on mixing. G^M may be expressed as

$$G^M = \sum_i x_i (\mu_i - \mu_i^0). \tag{I.11}$$

For the ideal solution, the free energy of mixing G^{IM} is given by

$$G^{IM} = RT \sum_i (x_i \ln x_i) \tag{I.12}$$

In (I.11) we have the quantity $(\mu_i - \mu_i^0)$ which represents the property change of a mole of component i as a result of mixing at constant P and T.

c) Non-Ideal Solutions and Excess Functions of Mixing

The relation between the chemical potential of a component i and its activity a_i in a solution is given by

$$\mu_i = \mu_i^0 + RT \ln a_i. \tag{I.13}$$

Ideal solution is the limiting case when a_i is equal to the mole fraction x_i. In all other cases, the a_i may be related to x_i by

$$a_i = f_i x_i \tag{I.14}$$

where f_i is the activity coefficient of the component i in the solution.

The molar free energy of mixing G^M for a solution is given by

$$\begin{aligned} G^M &= RT \sum_i (x_i \ln a_i) \\ &= RT \sum_i (x_i \ln x_i) + RT \sum (x_i \ln f_i) \\ &= G^{IM} + G^{EM} \end{aligned} \tag{I.15}$$

where G^{EM} is $RT \sum (x_i \ln f_i)$ and is the excess free energy of mixing. An excess property is defined as the difference between the actual property of the solution and the property of an ideal solution (see PRIGOGINE and DEFAY, 1954). The different excess functions of mixing may be expressed as

(a) Excess entropy:

$$\begin{aligned} S^{EM} &= -\frac{\partial G^{EM}}{\partial T} \\ &= -RT \left(x_1 \frac{\partial \ln f_1}{\partial T} + x_2 \frac{\partial \ln f_2}{\partial T} \right) - R(x_1 \ln f_1 + x_2 \ln f_2). \end{aligned} \tag{I.16}$$

(b) Excess enthalpy:

$$\begin{aligned} H^{EM} &= -T^2 \frac{\partial \left(\dfrac{G^{EM}}{T} \right)}{\partial T} \\ &= -RT^2 \left(x_1 \frac{\partial \ln f_1}{\partial T} + x_2 \frac{\partial \ln f_2}{\partial T} \right). \end{aligned} \tag{I.17}$$

(c) Excess volume:

$$\begin{aligned} V^{EM} &= \frac{\partial G^{EM}}{\partial P} \\ &= RT \left(x_1 \frac{\partial \ln f_1}{\partial P} + x_2 \frac{\partial \ln f_2}{\partial P} \right). \end{aligned} \tag{I.18}$$

Gibbs-Duhem equations are useful in relating activity or activity-coefficient to the mole fractions. These equations are

$$x_1 \left(\frac{\partial \ln a_1}{\partial x_2} \right) + x_2 \left(\frac{\partial \ln a_2}{\partial x_2} \right) = 0, \tag{I.19}$$

$$x_1 \left(\frac{\partial \ln f_1}{\partial x_2} \right) + x_2 \left(\frac{\partial \ln f_2}{\partial x_2} \right) = 0. \tag{I.20}$$

If the activity or activity coefficient of one of the components is known as a function of the mole fraction, then the activity or activity coefficient of the other function can be calculated by an integration as below (see WAGNER, 1952)

$$\ln f_1 = - \int\limits_{0}^{x_2} \frac{x_2}{1 - x_2} \frac{\partial \ln f_2}{\partial x_2} dx_2. \tag{I.21}$$

To facilitate the integration, $\ln f_2$ may be considered as the independent variable, thus determining x_2 and $1 - x_2$. At the lower integration limit in Eq.(I.21), $\ln f_2^0 = \ln f_2$ for $x_2 = 0$. We have

$$\ln f_1 = - \int\limits_{\ln f_2^0}^{\ln f_2(x_2)} \frac{x_2}{1 - x_2} d \ln f_2. \tag{I.22}$$

WAGNER (1952) suggests integrating Eq.(I.22) by parts:

$$\ln f_1 = \int\limits_{0}^{x_2} \frac{\ln f_2}{(1 - x_2)^2} dx_2 - \frac{x_2}{1 - x_2} \ln f_2. \tag{I.23}$$

The integral in Eq.(I.23) can be evaluated by graphical means if $\ln f_2$ is nearly proportional to $(1 - x_2)^2$ as in the simple mixture model of GUGGENHEIM discussed later.

d) Chemical Potential and Activity of a Component in a Mineral

In a binary solution α, whose composition is $(A, B)M$ where M may represent the anion group or the silicate framework, and A and B the cations which substitute for each other, there is a choice between adopting the cations A and B as components or the end member molecules AM and BM. As noted before, under certain conditions, the mole fractions may be calculated as

$$x_{A-\alpha} = \frac{A}{A + B} \quad \text{or} \quad x_{AM-\alpha} = \frac{AM}{AM + BM}$$

and these could be considered equivalent. We may write for chemical potentials

$$\mu_{A-\alpha} = \mu^0_{A-AM} + RT \ln x_{A-\alpha} f_{A-\alpha} \qquad (I.24)$$

or

$$\mu_{AM-\alpha} = \mu^0_{AM} + RT \ln x_{AM-\alpha} f_{AM-\alpha} \qquad (I.25)$$

where μ^0_{A-AM} and μ^0_{AM} are chemical potentials of A and AM in a standard state. The standard state AM is well defined but the standard state with reference to cation A needs definition. In orthopyroxene, we may refer to the chemical potential of Mg^{2+} in pure $(Mg, Mg)Si_2O_6$. The Gibbs free energy for the pure end member $(Mg, Mg)Si_2O_6$ is defined and measurable experimentally but the meaning of free energy of Mg^{2+} in pure orthoenstatite is little understood and experimental methods remain to be developed for its measurement.

However in many cases, where we are not concerned with the measured values of the potentials, the definition of chemical potential of a cation in a crystalline solution is not only permissible but also useful. KRETZ (1961) defines the chemical potential of Mg in orthopyroxene as

$$\mu_{Mg} = \left(\frac{\partial G}{\partial n_{Mg}} \right)_{P,T}, n_{Fe}, n_{Si}, n_O$$

where n is the number of cations in the formula.

In many crystalline solutions, when their compositions are expressed in the simplest form, there are two or more cations in one mole. Examples are olivine $(Fe, Mg)_2SiO_4$ and garnet $(Fe, Mg)_3Al_2Si_3O_{12}$. The chemical potential of a component using "molecular" model is expressed as:

$$\mu_{Fa-ol} = \mu^0_{Fa} + RT \ln x_{Fa-ol} f_{Fa-ol} \qquad (I.26)$$

$$\mu_{Alm-gar} = \mu^0_{Alm} + RT \ln x_{Alm-gar} f_{Alm-gar} \qquad (I.27)$$

where the μ^0_{Fa} and μ^0_{Alm} are the chemical potentials of the pure end members fayalite (Fe_2SiO_4) and almandine $(Fe_3Al_2Si_3O_{12})$, and μ_{Fa-ol} and $\mu_{Alm-gar}$ are the chemical potentials of fayalite and garnet in the solution. Generally, however, the experimentally determined thermodynamic quantities would be more in accord with the theory, if the "ionic" model is used (see BRADLEY, 1962, HOWARD DAY, personal communication). Thus in olivine and garnet, we have

$$\mu_{Fa-ol} = \mu^0_{Fa} + 2RT\ x_{Fa-ol} f_{Fa-ol} ,$$

$$\mu_{Alm-gar} = \mu^0_{Alm} + 3RT \ln x_{Alm-gar} f_{Alm-gar} .$$

It may be desirable to consider the chemical formula on one cation basis, i.e. we consider olivine as $(Fe, Mg)Si_{0.5}O_2$ and garnet as

$(Mg, Fe)Al_{2/3}SiO_4$. In these cases we may write

$$\mu_{\text{Fa-ol}} = 1/2\ \mu^0_{\text{Fa}} + RT\ \ln x_{\text{Fa-ol}} f_{\text{Fa-ol}} \tag{I.28}$$

and

$$\mu_{\text{Alm-gar}} = 1/3\ \mu^0_{\text{Alm}} + RT\ \ln x_{\text{Alm-gar}} f_{\text{Alm-gar}}. \tag{I.29}$$

In the above examples the mole fractions x are the same quantities in both the molecular and ionic models. They may be different in other minerals such as plagioclase where $x_{Ab}(Ab/Ab + An,\ Ab = NaAlSi_3O_8,\ An = CaAl_2Si_2O_8)$ is a different quantity from $x_{Na}(Na/Na + K)$.

The activities of the components will be the primary concern of this monograph. For a binary ideal solution, the activity is equal to its mole fraction. In olivine the activity of the fayalite (Fa) is

$$a_{\text{Fa}} = x^2_{\text{Fa-ol}} \tag{I.30}$$

or for Fe^{2+},

$$a_{\text{Fe}} = x^2_{\text{Fe-ol}}. \tag{I.31}$$

Similarly in garnet for Almandine (Alm) we have,

$$a_{\text{Alm}} = x^3_{\text{Alm-gar}} \tag{I.32}$$

and

$$a_{\text{Fe}} = x^3_{\text{Fe-gar}} \tag{I.33}$$

Again we may consider the reaction on a one-cation basis — i.e. consider olivine as $(Fe, Mg)Si_{0.5}O_2$, etc. This is particularly useful in treating ion-exchange equilibria. Activity of a cation is then equal to its mole fraction.

II. Thermodynamic Models
for Crystalline Solutions

Composition of coexisting minerals occurring in rocks or in experiments are the main source of data to be used in obtaining the information on the thermodynamic behaviour of silicate crystalline solutions. It is, therefore, necessary to use certain solution models to relate the observed compositional variables to the thermodynamic functions of mixing. Although such models are based on specific statistical theory and employ certain assumptions regarding the molecular forces in the crystalline lattice, the equations obtained for the thermodynamic functions of mixing are mathematically equivalent to those of other mathematical models which are not bound to any special physical interpretation. The regular solution model of GUGGENHEIM (1952) is considered in detail in this chapter.

1. Regular Solution Model

a) Zeroth Approximation

The excess free energy of mixing G^{EM} in a regular solution with the zeroth approximation, i.e., the approximation of complete disorder, is given by

$$G^{EM} = x_A x_B W', \tag{II.1}$$

where A and B are components of a solution (A, B) M and W' is equal to Nw, N being Avogadro's number. W' is often referred to as the interchange energy. Regular solutions are very important in this work; therefore, the parameter w will be briefly discussed. A simplified account of this parameter is presented by DENBIGH (1965). It is assumed that the cations A and B are of roughly the same size and can be interchanged between lattice sites without change of lattice structure and without change in the lattice vibrations. There is interaction between A and B, given by the energy w:

$$2w = 2w_{AB} - w_{AA} - w_{BB} \tag{II.2}$$

where w_{AA} is the increase in potential energy when a pair of A ions are brought together from infinite distance to their equilibrium separation in the solution. w_{AB} and w_{BB} are similarly defined. In spite of the interaction energy, it is assumed that the mixing of A and B is random. This means that the entropy of mixing is the same as that for an ideal solution and deviations are expressed entirely in terms of the heat of mixing.

The thermodynamic equations for the regular solution model with zeroth approximation are

$$G^{EM} = H^{EM} = x_A x_B W'$$ (II.3)

and

$$S^{EM} = 0.$$ (II.4)

The interchange energy W' is assumed to be independent of P and T. Because the excess entropy of mixing is zero according to this model, the predictions of the values of G^{EM} and the heat of mixing H^{EM}, which may often be different from G^{EM}, are not satisfactory.

b) Simple Mixture Model

In the regular solution model W' is supposed to be independent of T and P. In GUGGENHEIM's (1967) latest version of the lattice theory, W' may be treated as an adjustable constant required to fit the experimental data to the model. Such an energy parameter with a symbol W may be called a cooperative free energy. $2W$ is in a sense the free energy increase in the whole system when an A–A pair and a B–B pair are converted into two A–B pairs. It is expected that if W is fitted to the free energy data at each temperature, the large errors usually found in the predictions of G^{EM} and H^{EM} with composition may be at least partly eliminated. For a random mixing approximation, the various excess functions are given by

$$G^{EM} = x_A x_B W,$$ (II.5)

$$-S^{EM} = x_A x_B \frac{\partial W}{\partial T},$$ (II.6)

and

$$H^{EM} = x_A x_B \left(W - T \frac{\partial W}{\partial T} \right).$$ (II.7)

The activity coefficient is related to the mole fraction by

$$\ln f_A = \frac{W}{RT} x_B^2.$$ (II.8)

c) Quasi-Chemical Model

The main assumptions required for this model are similar to those of the regular solution model in the preceding sections. Only the configurational partition function of the solution contributes to the thermodynamic excess functions. The intermolecular forces are central and short range, and therefore the internal energy at $0\,°K$ may be obtained by an addition of the pair potentials. The assumption of complete randomness is not required here. Therefore any differences found in the calculated values of the excess functions of mixing by the zeroth approximation and by the quasi-chemical approximation are the result of ordering considered in the latter.

In binary solutions for which the two components A and B are of similar size, the activity coefficients are given by the equations

$$f_A = \left(\frac{\beta + 1 - 2x_B}{x_A(\beta + 1)}\right)^{z/2} \tag{II.9}$$

and

$$f_B = \left(\frac{\beta - 1 + 2x_B}{x_A(\beta + 1)}\right)^{z/2} \tag{II.10}$$

where z is the coordination number and β is given by

$$\beta = [1 + 4x_A x_B(e^{2W/zRT} - 1)]^{1/2}. \tag{II.11}$$

$\beta = 1$ for a perfectly random mixture. If $\beta > 1$, a tendency for clustering (positive deviation from ideal solution) exists, and if $\beta < 1$, a tendency for compound formation exists (negative deviation from ideal solution).

$$G^{EM} = \frac{1}{2} zRT \left[x_A \ln \frac{\beta + 1 - 2x_B}{x_A(\beta + 1)} + x_B \ln \frac{\beta - 1 + 2x_B}{x_A(\beta + 1)}\right] \tag{II.12}$$

and

$$H^{EM} = \frac{2}{\beta + 1} x_A x_B z \left(W - T\frac{\partial W}{\partial T}\right). \tag{II.13}$$

The various equations of the quasi-chemical approximation may be expanded as a power series in $2W/zRT$:

$$G^{EM} = Wx_A x_B \left[1 - \frac{1}{2}\left(\frac{2W}{zRT}\right) x_A x_B - \frac{1}{6}\left(\frac{2W}{zRT}\right)^2 x_A x_B(x_A - x_B)^2 + \cdots\right]$$

and
$$\tag{II.14}$$

$$f_A = Wx_B^2 \left[1 + \frac{1}{2}\left(\frac{2W}{zRT}\right) x_A(1 - 3x_B) + \frac{1}{6}\left(\frac{2W}{zRT}\right)^2\right.$$

$$\tag{II.15}$$

$$\left. \cdot x_A(1 - 11x_B + 28x_B^2 - 20x_B^3) + \cdots\right]$$

f_B may be obtained by replacing A by B in Eq. (II.15).

For molecules that are not very similar in size, a contact factor must be included (GUGGENHEIM, 1952, p. 186) in these equations to take the size differences into account. The contact factors may be found roughly proportional to the molar volumes or ionic radii. The activity coefficients are given by

$$f_A = \left[1 + \frac{\varphi_B(\beta - 1)}{\varphi_A(\beta + 1)}\right]^{z\,q_A/2} \tag{II.16}$$

and

$$f_B = \left[1 + \frac{\varphi_A(\beta - 1)}{\varphi_B(\beta + 1)}\right]^{z\,q_B/2} \tag{II.17}$$

where q_A and q_B are contact factors related to the contact fractions φ_A and φ_B and the mole fractions x_A and x_B by

$$\varphi_A = \frac{x_A q_A}{x_A q_A + x_B q_B} \tag{II.18a}$$

and

$$\varphi_B = \frac{x_B q_B}{x_A q_A + x_B q_B}. \tag{II.18b}$$

For more details on the derivation and significance of the constants q_A and q_B and the fractions φ_A and φ_B, reference may be made to GUGGEN-HEIM (1952, p. 186) and KING (1969, p. 488). β in Eqs. (II.16) and (II.17) is obtained by replacing x_A and x_B by φ_A and φ_B, respectively, in Eq. (II.11).

The other excess functions are given by

$$G^{EM} = \frac{1}{2}z\,R\,T\left\{x_A q_A \ln\left[1 + \frac{\varphi_B(\beta - 1)}{\varphi_A(\beta + 1)}\right] + x_B q_B \ln\left[1 + \frac{\varphi_A(\beta - 1)}{\varphi_B(\beta + 1)}\right]\right\} \tag{II.19}$$

and

$$H^{EM} = \frac{2 x_A x_B q_A q_B}{(x_A q_A + x_B q_B)(\beta + 1)}\left(W - T\frac{\partial W}{\partial T}\right). \tag{II.20}$$

S^{EM} can be obtained by the standard equation

$$G^{EM} = H^{EM} - T S^{EM}.$$

2. General Relations for Binary, Ternary, and Quaternary Nonideal Crystalline Solutions

Excess functions in nonideal solutions may conveniently be expressed by a power series in the mole fractions. GUGGENHEIM (1937) suggested that G^{EM} can be expressed as a polynomial in x as

$$G^{EM} = x_A x_B [A_0 + A_1(x_A - x_B) + A_2(x_A - x_B)^2 + \cdots], \tag{II.21}$$

where A_0, A_1, and A_2 are constants. When odd terms in Eq. (II.21) vanish, the solution becomes symmetric in the sense that the excess functions of mixing associated with each end member vary symmetrically with respect to each other. If A_2 and other higher terms are also zero, the simple mixture model with A_0 as the energy constant W in Eq. (II.5) results. The expressions for the activity coefficients are obtained from

$$RT \ln f_A = G^{EM} + x_B \frac{\partial G^{EM}}{\partial x_A}$$

$$= x_B^2 [A_0 + A_1(3x_A - x_B) + A_2(x_A - x_B)(5x_A - x_B) + \cdots]$$

(II.22)

and

$$RT \ln f_B = G^{EM} - x_A \frac{\partial G^{EM}}{\partial x_B}$$

$$= x_A^2 [A_0 - A_1(3x_B - x_A) + A_2(x_B - x_A)(5x_B - x_A) + \cdots].$$

(II.23)

Equations for other excess functions of mixing may be derived from Eq. (II.21):

$$-S^{EM} = x_A x_B \left[\frac{\partial A_0}{\partial T} + \left(\frac{\partial A_1}{\partial T}\right)(x_A - x_B) + \left(\frac{\partial A_2}{\partial T}\right)(x_A - x_B)^2 + \cdots \right]$$ (II.24)

and

$$H^{EM} = x_A x_B \left\{ A_0 - T\left(\frac{\partial A_0}{\partial T}\right) + \left[A_1 - T\left(\frac{\partial A_1}{\partial T}\right)\right](x_A - x_B) \right.$$

$$\left. + \left[A_2 - T\left(\frac{\partial A_2}{\partial T}\right)\right](x_A - x_B)^2 + \cdots .$$

(II.25)

In mineralogical literature, there are different systems of equations used for the thermodynamic functions of mixing. These systems of equations are often simply related and may be derived from the general relation (II.21) (see VAN NESS, 1964):

$$G^{EM} = x_A x_B [A_0 + A_1(x_A - x_B) + A_2(x_A - x_B)^2 + \cdots].$$

Let us consider the following special cases.

1. If $A_0 = A_1 = A_2 = \cdots 0$, then G^{EM} also $= 0$. In this case, therefore, the solution is ideal.

2. If $A_0 \neq 0$ but $A_1 = A_2 = \cdots = 0$, then

$$G^{EM} = x_A x_B A_0,$$

$$RT \ln f_A = x_B^2 A_0 \text{ and}$$

$$RT \ln f_B = x_A^2 A_0.$$

These relations are similar to those for regular solution or simple mixture of GUGGENHEIM if we change the symbol A_0 to W' or W. Such relations were first proposed by PORTER (1921) and may also be referred to as Porter equations.

3. When $A_0 \neq 0$, $A_1 \neq 0$ but $A_2 = A_3 = \cdots = 0$, then

$$G^{EM} = x_A x_B [A_0 + A_1(x_A - x_B)],$$

$$RT \ln f_A = x_B^2 [A_0 + A_1(3x_A - x_B)],$$

$$RT \ln f_B = x_A^2 [A_0 - A_1(3x_B - x_A)].$$

These are GUGGENHEIM's two constant equations and will be commonly used in this book. These two constant equations are similar to Margules type used by THOMPSON (1967) and THOMPSON and WALDBAUM (1969a, b). These are

$$G^{EM} = x_A x_B (W_{G_1} x_B + W_{G_2} x_A), \tag{II.26}$$

$$RT \ln f_A = x_B^2 [W_{G_1} + 2x_A(W_{G_2} - W_{G_1})], \tag{II.27}$$

$$RT \ln f_B = x_A^2 [W_{G_2} + 2x_B(W_{G_1} - W_{G_2})]. \tag{II.28}$$

where W_G is Margules parameter.

The equivalence of the two sets of equations may be shown as below. THOMPSON's (1967) equation for G^{EM} is

$$G^{EM} = x_A G_2 + x_B G_1 \tag{II.29}$$

where

$$G_2 = x_A x_B W_{G_2} \tag{II.30}$$

and

$$G_1 = x_A x_B W_{G_1}. \tag{II.31}$$

This is as if the crystalline solution is composed of x_A moles of a simple mixture with W_{G_2} and x_B moles of another simple mixture with W_{G_1}. Then,

$$\begin{aligned} G^{EM} &= (W_{G_1} x_B + W_{G_2} x_A) x_A x_B \\ &= W_{G_1}[(1 - x_A) + W_{G_2}(1 - x_B)] x_A x_B \tag{II.32} \\ &= W_{G_1}\left(\frac{2 - 2x_A}{2}\right) + W_{G_2}\left(\frac{2 - 2x_B}{2}\right) x_A x_B. \end{aligned}$$

Substituting $1 = x_A + x_B$ into Eq. (II.32)

$$\begin{aligned} G^{EM} &= \left[\frac{W_{G_1}}{2}(1 - x_A + x_B) + \frac{W_{G_2}}{2}(1 + x_A - x_B)\right] x_A x_B \\ &= \left[\frac{W_{G_2} + W_{G_1}}{2} + \frac{W_{G_2} - W_{G_1}}{2}(x_A - x_B)\right] x_A x_B \end{aligned} \tag{II.33}$$

which is of the same form as GUGGENHEIM's equation with two constants A_0 and A_1.

Therefore,

$$A_0 = \frac{W_{G_2} + W_{G_1}}{2} \tag{II.34a}$$

and

$$A_1 = \frac{W_{G_2} - W_{G_1}}{2}. \tag{II.34b}$$

A_0/RT and A_1/RT would correspond to the notation B_G and C_G, respectively, used by THOMPSON (1967) following SCATCHARD and HAMER (1935).

For the activity coefficient,

$$RT \ln f_A = (x_B)^2 \left[A_0 + A_1(3x_A - x_B) + \cdots \right]. \tag{II.22}$$

Substitution of Eq. (II.34) into Eq. (II.22) gives

$$
\begin{aligned}
RT \ln f_A &= (x_B)^2 \left[\frac{W_{G_2} + W_{G_1}}{2} + \frac{W_{G_2} - W_{G_1}}{2}(x_A - x_B + 2x_A) \right] \\
&= x_B^2 \left\{ \left[\frac{W_{G_2} + W_{G_1}}{2} + \frac{W_{G_2} - W_{G_1}}{2}(x_A - x_B) \right] + (W_{G_2} - W_{G_1})x_A \right\} \\
&= x_B^2 [(W_{G_1} x_B + W_{G_2} x_A) + (W_{G_2} x_A - W_{G_1} x_A)] \tag{II.27} \\
&= x_B^2 [W_{G_1}(x_B - x_A) + 2W_{G_2} x_A] \\
&= x_B^2 [W_{G_1}(1 - 2x_A) + 2W_{G_2} x_A] \\
&= x_B^2 [W_{G_1} + 2x_A(W_{G_2} - W_{G_1})]
\end{aligned}
$$

which is of the same form as used by THOMPSON (1967).

An alternative to the expression (II.21) for G^{EM} may be written as

$$\frac{1}{G^{EM}} = x_A x_B [A_0 + A_1(x_A - x_B) + A_2(x_A - x_B)^2 + \cdots] \tag{II.35}$$

Activity coefficients are given by

$$RT \ln f_A = \frac{(x_B)^2 [A_0 - A_1 - A_2(2x_B - 1)(2x_B - 3) + \cdots]}{[A_0 - A_1(2x_B - 1) + A_2(2x_B - 1)^2 - \cdots]}, \tag{II.36}$$

$$RT \ln f_B = \frac{(x_A)^2 [A_0 + A_1 - A_2(2x_A - 1)(2x_A - 3) - \cdots]}{[A_0 + A_1(2x_A - 1) + A_2(2x_A - 1)^2 + \cdots]}. \tag{II.37}$$

When $A_0 \neq 0$, $A_1 \neq 0$, $A_2 = A_3 = \cdots = 0$, Eqs. (II.36) and (II.37) reduce to the well known van Laar equations

$$RT \ln f_A = \frac{x_B^2 (A_0 - A_1)}{[A_0 - A_1 (2x_B - 1)]^2} \tag{II.38}$$

and

$$RT \ln f_B = \frac{x_A^2 (A_0 + A_1)}{[A_0 + A_1 (2x_A - 1)]^2} \tag{II.39}$$

van Laar's equations resulted from a theory based on the van der Waals equation of state. This theory is probably incorrect, but van Laar's equations continue to be useful for representing the activity-composition relation. For many chemical systems, van Laar's equation provides a better representation of the data than the Margules two-constant equation. The relative merits of these two equations are discussed by CARLSON and COLBURN (1942).

For a ternary solution WOHL (1953, quoted in KING, 1969) derived the following expression for activity coefficient

$$\begin{aligned}
\ln f_A = &\, x_B^2 [E_{AB} + 2x_A (E_{BA} - E_{AB})] + x_C^2 [E_{AC} + 2x_A \\
&\, (E_{CA} - E_{AC})] + x_B x_C [1/2 (E_{BA} + E_{AB} + E_{CA} \\
&\, + E_{AC} - E_{BC} - E_{CB}) + x_A (E_{BA} - E_{AB} + E_{CA} \\
&\, - E_{AC}) + (x_B - x_C)(E_{BC} - E_{CB}) - (1 - 2x_A) C].
\end{aligned} \tag{II.40}$$

An analogous equation for $\ln f_B$ is obtained by replacing subscript A by B, B by C and C by A in Eq. (II.40). The equation for $\ln f_C$ is obtained by replacing subscript A by C, C by B and B by A in Eq. (II.40). The six E's are binary constants, two each for the three binary systems, which are equivalent to Margules parameters. These equations, therefore, may be referred to as the ternary Margules equations. C is a ternary constant. Equation (II.40) will be used in the next chapter, where the effect of changing the ternary constant C in some hypothetical cases will be considered. By neglecting the ternary constant, Eq. (II.40) or the following equation may be used to predict ternary equilibrium data from binary data

$$G^{EM} = \sum_{ij} x_i x_j [B_{ij} + C_{ij}(x_i - x_j) + D_{ij}(x_i - x_j)^2 + \cdots] \tag{II.41}$$

where B, C and D correspond to A_0, A_1 and A_2.

III. Thermodynamic Stability of a Solution

1. Critical Mixing

For binary or multi-component crystalline solutions, the stability is determined by the diffusion processes. Above a certain temperature two end member crystals may form a complete crystalline solution series. At lower temperatures, the solution may become unstable in a certain range of the composition. Figure 1 shows a miscibility gap in a binary crystalline solution (A, B)M. Above T_c, the critical temperature of unmixing, the solution is continuous between AM and BM and irrespective of whether AM and BM are structurally similar or dissimilar, one can pass from an AM rich phase to a BM rich phase without any observable break or transition point. Below T_c there are two coexisting phases both crystalline solutions one rich in AM and the other rich in BM. At the *critical solution point* C the two phases become identical.

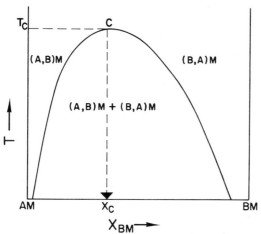

Fig. 1. Miscibility gap in a binary solution (A, B)M. C is the critical solution point, x_c the critical composition and T_c the critical temperature

2. General Conditions

The conditions for critical mixing in terms of G^M and the mole fraction are

$$\partial^2 G^M / \partial x^2 = 0 \qquad (III.1)$$

and
$$\partial^3 G^M/\partial x^3 = 0. \tag{III.2}$$

These may be expressed in terms of G^{EM} as

$$\partial^2 G^{EM}/\partial x^2 = -RT/x(1-x) \tag{III.3}$$

and
$$\partial^3 G^{EM}/\partial x^3 = -RT(2x-1)/x^2(1-x)^2. \tag{III.4}$$

a) Simple Mixture

For a simple mixture,

$$G^{EM} = x(1-x)W, \tag{III.5}$$

where
$$W = W(T, P).$$

By successive differentiation of Eq. (III.5),

$$\partial^2 G^{EM}/\partial x^2 = -2W \tag{III.6}$$

and
$$\partial^3 G^{EM}/\partial x^3 = 0. \tag{III.7}$$

By substituting Eqs. (III.3) and (III.4) into Eqs. (III.6) and (III.7) respectively,

$$-2W = -RT/x(1-x) \tag{III.8}$$

and
$$0 = RT(2x-1)/x^2(1-x)^2. \tag{III.9}$$

These give the critical composition when $x = 0.5$ and $2RT_c = W$.

b) General Nonideal Solution

For a binary solution that is not a symmetric solution,

$$G^{EM} = (1-x)[A_0 + A_1(2x-1) + A_2(2x-1)^2 + \cdots]. \tag{III.10}$$

Successive differentiation of Eq. (III.10) with respect to x gives

$$\frac{\partial^2 G^{EM}}{\partial x^2} = -2A_0 - 6A_1(2x-1) - A_2[10 - 48x(1-x)] \tag{III.11}$$

and
$$\frac{\partial^3 G^{EM}}{\partial x^3} = -12A_1 + 48A_2(1-2x). \tag{III.12}$$

Substitution of Eqs. (III.3) and (III.4) into Eqs. (III.11) and (III.12), respectively gives equations that are transcendental and cannot be solved without a computer program using an iteration method.

3. Spinodal Decomposition

Figure 2 shows chemical potential μ_A plotted against the mole fraction x_B in a crystalline solution $(A, B)M$ which unmixes into two crystalline solutions below the critical temperature. At C the two crystalline solutions become identical. It may, therefore, be regarded that the two portions of curve 3 (Fig. 3) as two segments of the continuous curve abmncd. The portions bm and cn are metastable extensions of the two segments a and d respectively. This is essentially a kinetic phenomenon and is of interest in phase-separation theories in crystalline solutions. The states between m and n (Fig. 3) are unstable and are characterized by

$$\frac{\partial \mu_A}{\partial x_B} = 0 \qquad\qquad \text{(III.14)}$$

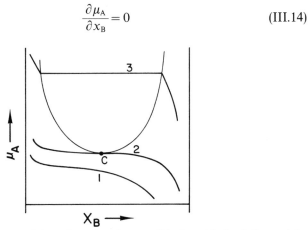

Fig. 2. Chemical potential μ_A plotted against x_B. C is the critical solution point

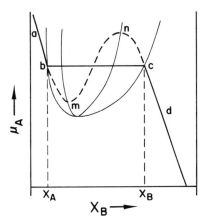

Fig. 3. The chemical potential curve shown as continuous abmncd. The portions bm and cn are metastable and mn is unstable

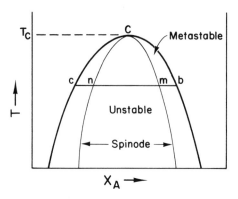

Fig. 4. Metastable and unstable regions of a miscibility gap

In terms of Gibbs free energy per mole, $\partial^2 G/\partial x^2$ is positive between bm and equals 0 at m; is negative from m to n, and equals 0 at n; and is positive from c to d. Figure 4 shows the unstable and metastable regions. The boundary between the two regions is denoted as the spinode. Evidently for a two constant equation for the free energy of mixing, the spinodal is given by

$$\frac{\partial^2 G}{\partial x_B^2} = 0 = \frac{RT}{x_A x_B} - 2A_0 + 6A_1(1 - 2x_A) \tag{III.15}$$

which is equivalent to the equation with Margules parameters given by THOMPSON (1967).

4. Critical Mixing in Quasi-Chemical Approximation

Following GREEN (1970) the excess chemical potential may be expressed as

$$\frac{\mu_1^E}{RT} = \frac{zq_1}{2} \ln\left[1 + \frac{\varphi_2(\beta - 1)}{\varphi_1(\beta + 1)}\right] \tag{III.16}$$

where μ_1^E is the excess chemical potential of component 1 in solution; other symbols are as explained in Chapter II.

Differentiating (III.16) with respect to x_2, we have

$$\frac{1}{RT}\left(\frac{\partial \mu_1}{\partial x_2}\right) T_s = 0 = \frac{1}{x_1}\left[-1 + \frac{zq_1 q_2(\beta - 1)}{2(x_1 q_1 + x_2 q_2)\beta}\right]. \tag{III.17}$$

The spinodal curve is given by

$$\beta(T_s) = \frac{q_1 q_2 z}{q_1 q_2 z - 2(x_1 q_1 + x_2 q_2)} \qquad \text{(III.18)}$$

where T_s is the maximum in the spinodal such that $dT_s/dx_2 = 0$. Differentiating again with respect to x_2, setting $dT_s/dx = 0$ and combining (III.18) with (II.11), we obtain

$$\left(\frac{\partial \beta}{\partial x_2}\right) T_s = \frac{(\beta^2 - 1)(\varphi_1 - \varphi_2)}{2\beta x_1 x_2} = \frac{2q_1 q_2 z(q_2 - q_1)}{[q_1 q_2 z - 2(x_1 q_1 + x_2 q_2)]^2} \qquad \text{(III.19)}$$

which is rearranged using (11) of Chapter II to yield

$$[x_2 q_2 - x_1 q_1][q_1 q_2 z - 2(x_1 q_1 + x_2 q_2)]$$
$$+ [(q_2 - q_1)q_1^2 q_2^2 z^2 x_1 x_2] = 0 \qquad \text{(III.20)}$$

φ_1 and φ_2 are defined in Chapter II. The critical solution composition may be determined from (III.20) by using a numerical method. The critical temperature is then found from Eq.(III.18) using the values of critical solution compositions x_{1C} and x_{2C}.

5. Immiscibility in Ternary Crystalline Solutions

A critical point in ternary solution is characterized by the equations (see PRIGOGINE and DEFAY, 1954):

$$\left(\frac{\partial \mu_2}{\partial x_2}\right) T, P, \mu_1, x_3 = 0$$

$$\left(\frac{\partial^2 \mu_2}{\partial x_2^2}\right) T, P, \mu_1, x_3 = 0 \qquad \text{(III.21)}$$

$$\left(\frac{\partial^3 \mu_2}{\partial x_2^3}\right) T, P, \mu_1, x_3 \neq 0.$$

Figure 5 shows the molar Gibbs free energy surface for a ternary solution at a constant P and T. The line acb represents the spinodal curve and is given by

$$\frac{\partial^2 G}{\partial x_2^2} \frac{\partial^2 G}{\partial x_3^2} - \left(\frac{\partial^2 G}{\partial x_2 \partial x_3}\right)^2 = 0 \qquad \text{(III.22)}$$

M and N are coexisting phases and the line ABC is the coexistence or binodal curve. At M and N the tangent planes to the free energy surface are coincident which follow from the requirement that

$$\mu_{1-M} = \mu_{1-N}, \mu_{2-M} = \mu_{2-N}, \mu_{3-M} = \mu_{3-N}.$$

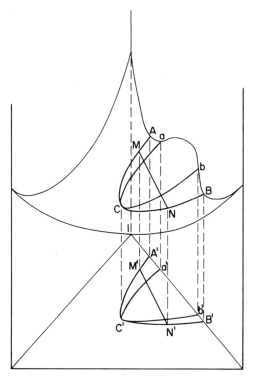

Fig. 5. Molar Gibbs free energy surface in a ternary solution. ACB and aCb are
the binodal and spinodal curves respectively

C is the critical point and the tangent at this point is the limit to which
the tie lines tend as C is approached.

In what follows, it is attempted to use two of the solution models to
define the excess free energy of mixing and the activities of components
in ternary solutions. These functions of mixing are then used to calculate
the composition of the coexisting phases at different temperatures.

6. Formation of Miscibility Gaps in a Ternary Simple Mixture

Consider a ternary simple mixture with components 1, 2, and 3. W for
the three binary systems are W_{12}, W_{13}, and W_{23}. The chemical potentials
of the components in the solution are given by

$$\mu_i = \mu_i^0(T, P) + RT \ln x_i + RT \ln f_i, \qquad \text{(III.23)}$$

where $RT \ln f$ may be expanded in terms of x and W as follows:

$$RT \ln f_1 = (x_2)^2 W_{12} + (x_3)^2 W_{13} + x_2 x_3 (W_{12} - W_{23} + W_{13}), \qquad \text{(III.24a)}$$

$$RT \ln f_2 = (x_3)^2 W_{23} + (x_1)^2 W_{12} + x_3 x_1 (W_{23} - W_{13} + W_{12}), \qquad \text{(III.24b)}$$

and

$$RT \ln f_3 = (x_1)^2 W_{13} + (x_2)^2 W_{23} + x_1 x_2 (W_{13} - W_{12} + W_{23}). \qquad \text{(III.24c)}$$

At equilibrium in the two separated coexisting phases α and β,

$$\mu_{1-\alpha}(x_{2-\alpha}, \ x_{3-\alpha}, T) - \mu_{1-\beta}(x_{2-\beta}, \ x_{3-\beta}, T) = 0 \qquad \text{(III.25)}$$

μ_2 and μ_3 are similarly defined. Substituting Eqs. (III.23) and (III.24) into Eq. (III.25) and rearranging (see KAUFMAN and BERNSTEIN, 1970, p. 226),

$$RT \ln(x_{1-\beta}/x_{1-\alpha}) + W_{12}[(x_{2-\beta})^2 - (x_{2-\alpha})^2] + W_{13}[(x_{3-\beta})^2 - (x_{3-\alpha})^2]$$
$$+ \Delta W(x_{2-\beta} x_{3-\beta} - x_{2-\alpha} x_{3-\alpha}) = 0 \qquad \text{(III.26a)}$$

$$RT \ln(x_{2-\beta}/x_{2-\alpha}) + W_{12}[(1 - x_{2-\beta})^2 - (1 - x_{2-\alpha})^2]$$
$$+ W_{13}[(x_{3-\beta})^2 - (x_{3-\alpha})^2] - \Delta W[x_{3-\beta}(1 - x_{2-\beta})$$
$$- x_{2-\alpha}(1 - x_{3-\alpha})] = 0 \qquad \text{(III.26b)}$$

and

$$RT \ln(x_{3-\beta}/x_{3-\alpha}) + W_{12}[(x_{2-\beta})^2 - (x_{2-\alpha})^2]$$
$$+ W_{13}[(1 - x_{3-\beta})^2 - (1 - x_{3-\alpha})^2] - \Delta W[x_{2-\beta}(1 - x_{3-\beta})$$
$$- x_{2-\alpha}(1 - x_{3-\alpha})] = 0 \qquad \text{(III.26c)}$$

where $\Delta W = W_{12} + W_{13} - W_{23}$.

With the help of Eqs. (III.26), if W_{12}, W_{23} and W_{13} are given, compositions of coexisting phases may be calculated and the miscibility gap may be plotted on a ternary diagram. However, first the compositions of the coexisting phases on three binary edges must be calculated.

In a binary solution, the miscibility gap can be calculated by finding the composition of the coexisting phases that together represent the minimum free energy of the system. This may be done graphically by the tangent method, i.e., by drawing a tangent through the two points representing the two minima in the plot of the free energy of mixing against composition. Alternately, the relations

$$\mu_{1-\alpha} = \mu_{1-\beta}$$

and

$$\mu_{2-\alpha} = \mu_{2-\beta}$$

may be considered. For the binary regular solution, there is a symmetric miscibility gap and therefore

$$x_{1-\alpha} + x_{2-\alpha} = 1$$

$$x_{1-\beta} + x_{2-\beta} = 1$$

$$x_{1-\alpha} = x_{1-\beta}$$

and

$$x_{2-\alpha} = x_{2-\beta} \, .$$

Therefore,

$$RT \ln(1 - x_1) + x_1^2 W = RT \ln(1 - x_2) + x_2^2 W \, . \qquad \text{(III.27)}$$

$$RT \ln x_1 + (1 - x_1)^2 W = RT \ln x_2 + (1 - x_2)^2 W \, . \qquad \text{(III.28)}$$

Substituting $x_2 = 1 - x_1$ into Eq. (III.27),

$$\frac{W}{RT} = \frac{1}{1 - 2x_1} \ln \frac{1 - x_1}{x_1} \, . \qquad \text{(III.29)}$$

Equation (III.29) may be solved by an iteration method to find the miscibility gaps on the binary edges in a triangular diagram.

A computer program to solve Eq. (III.16) numerically and the method to form a miscibility gap have been presented by KAUFMAN and BERNSTEIN (1970). Examples to illustrate the possible solutions of certain mineralogical problems are presented elsewhere.

7. Formation of Miscibility Gap in Asymmetric Ternary Solutions

Since most crystalline solutions show a miscibility gap on the binary join which is not a simple symmetric mixture, it is necessary to consider two constant equations for the free energy of mixing in the three constituent systems. In a ternary solution, the activity coefficient is given by Eq. (II.40)

$$
\begin{aligned}
\ln f_1 = {} & x_2^2 [E_{12} + 2x_1(E_{21} - E_{12})] + x_3^2 [E_{13} + 2x_1(E_{31} - E_{13})] \\
& + x_2 x_3 [1/2(E_{21} + E_{12} + E_{31} + E_{13} - E_{23} - E_{32}) \\
& + x_1(E_{21} - E_{12} + E_{31} - E_{13}) + (x_2 - x_3)(E_{23} - E_{32}) \\
& - (1 - 2x_1)C]
\end{aligned}
\qquad \text{(II.40)}
$$

where E_{ij}'s are constants similar to W_{G_1}, W_{G_2}, etc. of THOMPSON (1967) and C the ternary constant.

The mathematical criterion for forming a miscibility gap is, as before, the equilibration of the chemical potentials across the gap,

$$\mu_{1-\alpha}(x_{2-\alpha}, x_{3-\alpha}, T) = \mu_{1-\beta}(x_{2-\beta}, x_{3-\beta}, T) \qquad \text{(III.30)}$$

where 1, 2 and 3 are components and α and β are coexisting crystalline solutions. We have similar relations for μ_2 and μ_3. Substituting the values of $\ln f_1$ from Eq. (II.40) and of $\ln f_2$ and $\ln f_3$ from similar relations, we finally obtain the three equations as:

$$RT \ln(x_{3-\beta}/x_{3-\alpha}) + E_{12}(x_{1-\beta}^2 - x_{1-\alpha}^2) + 2(E_{21} - E_{12})$$

$$(x_{1-\beta}x_{3-\beta} - x_{1-\alpha}x_{3-\alpha}) + E_{13}(x_{2-\beta}^2 - x_{2-\alpha}^2) + 2(E_{31} - E_{13})$$

$$(x_{2-\beta}^2 x_{3-\beta} - x_{2-\alpha}^2 x_{3-\alpha}) + 1/2(E_{21} + E_{12} + E_{31} + E_{13} - E_{23} - E_{32})$$

$$(x_{1-\beta}x_{2-\beta} - x_{1-\alpha}x_{2-\alpha}) + (E_{21} - E_{12} + E_{31} - E_{13})(x_{1-\beta}x_{2-\beta}x_{3-\beta}$$

$$- x_{1-\alpha}x_{2-\alpha}x_{3-\alpha}) + (E_{23} - E_{32})(x_{1-\beta}^2 x_{2-\beta} - x_{1-\beta}x_{2-\beta}^2 \qquad \text{(III.31)}$$

$$- x_{1-\alpha}^2 x_{2-\alpha} + x_{1-\alpha}x_{2-\alpha}^2) - C(x_{1-\beta}x_{2-\beta} - 2x_{1-\beta}x_{2-\beta}x_{3-\beta}$$

$$+ x_{1-\alpha}x_{2-\alpha} - 2x_{1-\alpha}x_{2-\alpha}x_{3-\alpha}) = 0,$$

$$RT \ln(x_{1-\beta}/x_{1-\alpha}) + E_{23}(x_{2-\beta}^2 - x_{2-\alpha}^2) + 2(E_{32} - E_{23})$$

$$(x_{2-\beta}^2 x_{1-\beta} - x_{2-\alpha}^2 x_{1-\alpha}) + E_{21}\{(1 - x_{1-\beta} - x_{2-\beta})^2$$

$$- (1 - x_{1-\alpha} - x_{2-\alpha})^2\} + 2(E_{12} - E_{21})\{(1 - x_{1-\beta} - x_{2-\beta})^2 x_{1-\beta}$$

$$- (1 - x_{1-\alpha} - x_{2-\alpha})^2 x_{1-\alpha}\} + 1/2(E_{32} + E_{23} + E_{12} + E_{21} - E_{31} - E_{13})$$

$$\{x_{2-\beta}(1 - x_{1-\beta} - x_{2-\beta}) - x_{2-\alpha}(1 - x_{1-\alpha} - x_{2-\alpha})\}$$

$$+ (E_{32} - E_{23} + E_{12} - E_{21})\{x_{1-\beta}x_{2-\beta}(1 - x_{1-\beta} - x_{2-\beta}) \qquad \text{(III.32)}$$

$$- x_{1-\alpha}x_{2-\alpha}(1 - x_{1-\alpha} - x_{2-\alpha})\} + (E_{31} - E_{13})\{x_{2-\beta}^2$$

$$(1 - x_{1-\beta} - x_{2-\beta}) - x_{2-\beta}(1 - x_{1-\beta} - x_{2-\beta})^2 - x_{2-\alpha}^2(1 - x_{1-\alpha} - x_{2-\alpha})$$

$$+ x_{2-\alpha}(1 - x_{1-\alpha} - x_{2-\alpha})^2\} - C\{x_{2-\beta}(1 - x_{1-\beta} - x_{2-\beta})$$

$$- 2x_{2-\beta}x_{1-\beta}(1 - x_{1-\beta} - x_{2-\beta}) + x_{2-\alpha}(1 - x_{1-\alpha} - x_{2-\alpha})$$

$$- 2x_{2-\alpha}x_{1-\alpha}(1 - x_{1-\alpha} - x_{2-\alpha})\} = 0,$$

$$RT \ln(x_{2-\beta}/x_{2-\alpha}) + E_{31}[(1-x_{1-\beta}-x_{2-\beta})^2 - (1-x_{1-\alpha}-x_{2-\alpha})^2]$$
$$+ 2(E_{13} - E_{31})\{(1-x_{1-\beta}-x_{2-\beta})^2 x_{2-\beta} - (1-x_{1-\alpha}-x_{2-\alpha})^2 x_{2-\alpha}\}$$
$$+ E_{32}(x_{1-\beta}^2 - x_{1-\alpha}^2) + 2(E_{23} - E_{32})(x_{1-\beta}^2 x_{2-\beta} - x_{1-\alpha}^2 x_{2-\alpha})$$
$$+ 1/2(E_{13} + E_{31} + E_{23} + E_{32} - E_{12} - E_{21}) \tag{III.33}$$
$$\{(1-x_{1-\beta}-x_{2-\beta})x_{1-\beta} - (1-x_{1-\alpha}-x_{2-\alpha})x_{1-\alpha}\} + (E_{13} - E_{31} + E_{23}$$
$$- E_{32})\{x_{1-\beta}x_{2-\beta}(1-x_{1-\beta}-x_{2-\beta}) - x_{1-\alpha}x_{2-\alpha}(1-x_{1-\alpha}-x_{2-\alpha})\}$$
$$+ (E_{12} - E_{21})\{(1-x_{1-\beta}-x_{2-\beta})^2 x_{1-\beta} - (1-x_{1-\beta}-x_{2-\beta})x_{1-\beta}^2$$
$$- (1-x_{1-\alpha}-x_{2-\alpha})^2 x_{1-\alpha} + (1-x_{1-\alpha}-x_{2-\alpha})x_{1-\alpha}^2\}$$
$$- C\{x_{1-\beta}(1-x_{1-\beta}-x_{2-\beta}) - 2x_{1-\beta}x_{2-\beta}(1-x_{1-\beta}-x_{2-\beta}) + x_{1-\alpha}$$
$$(1-x_{1-\alpha}-x_{2-\alpha}) - 2x_{1-\alpha}x_{2-\alpha}(1-x_{1-\alpha}-x_{2-\alpha})\} = 0.$$

In the latter two equations, the relations

$$x_{1-\alpha} + x_{2-\alpha} + x_{3-\alpha} = 1$$
$$x_{1-\beta} + x_{2-\beta} + x_{3-\beta} = 1$$

have been used to eliminate x_3.

The binary miscibility gaps on each edge are first determined by solving the following two equations for x_1 and x_2

$$\ln x_{1-\alpha} + (1-x_{1-\alpha})^2 \{A_0 + A_1(4x_{1-\alpha} - 1)\}$$
$$= \ln x_{1-\beta} + (1-x_{1-\beta})^2 \{A_0 + A_1(4x_{1-\beta} - 1)\} \tag{III.34a}$$
$$\ln(1-x_{1-\alpha}) + x_{1-\alpha}^2 \{A_0 - A_1(3 - 4x_{1-\alpha})\}$$
$$= \ln(1-x_{1-\beta}) + x_{1-\beta}^2 \{A_0 - A_1(3 - 4x_{1-\beta})\} \tag{III.34b}$$

where $A_0 = (E_{12} + E_{21})/2$ and $A_1 = (E_{21} - E_{12})/2$.

The ternary solution is then followed by the Newton-Raphson iteration technique. The program for calculating the composition of coexisting phases is presented in the Appendix. Although information on binary crystalline silicate solutions is becoming increasingly available, the ternary constant C is difficult to evaluate. The influence of the constant C and composition of coexisting crystalline solutions is now discussed in some hypothetical ternary systems.

Figure 6 shows binodal curves in a hypothetical ternary solution $(A, B, C)M$. The constants E_{ij}'s and C are;

$$E_{AB} = 500, \ E_{BA} = 4000, \ E_{AC} = 2000, \ E_{CA} = 5000, \ E_{BC} = 300,$$
$$E_{CB} = 1000 \ \text{and} \ C = 0, \text{cal/mole}.$$

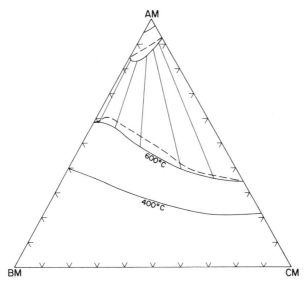

Fig. 6. Miscibility gaps in a ternary solution (A, B, C)M at 673 and 873 °K. The constants at both temperatures are $E_{AB} = 500$, $E_{BA} = 4000$, $E_{AC} = 2000$, $E_{CA} = 5000$, $E_{BC} = 300$, $E_{CB} = 1000$ and $C = 0$ cal/mol. The dashed curves indicate the change in the binodal at 873 °K by assuming $C = -1000$ cal/mol

These constants are a function of P and T but for illustration here, the same constants are used at different temperatures. Note that the tie lines rotate from the edge with the highest values of the E_{ij} constants to the edge with the next highest values of the constants.

The ternary constant C is varied in calculations represented by the dashed curve in Fig. 6. It is noted that moderate values of this constant (0 to 500 cal/mole) have little effect on the position of the binodal curve. High positive or negative value ($\sim$ 1000 cal/mole) change the miscibility gap significantly. The experimental values of C in several ternary organic solutions have been determined to be negligible. No such data on crystalline silicate solutions are available.

8. Immiscibility in Mineral Systems

Solution models discussed in Chapter II require a certain knowledge of the crystal structure of the endmember crystals and of the crystalline solutions. For many minerals such data are not available and the statistical units of the crystalline solutions are not well defined. A thermodynamic treatment of such systems may be made solely in terms of the macroscopic variables in the crystal as suggested later in Chapter V. Although quantitative data on solution properties for silicate minerals are lacking, it is possible to consider the petrologic consequences of immiscibility in minerals mainly based on the observations of mineral assemblages and composition of coexisting phases. In certain cases the thermodynamic properties of crystalline solutions may be determined quantitatively, as discussed before for binary solutions. In other cases, particularly for ternary solutions, we may be able to obtain qualitative to semi-quantitative information on solution properties.

a) Interpretation of Sequences of Mineral Assemblages

Following MUELLER (1964) immiscibility in binary and quasi-binary solutions may be used to explain certain sequences of mineral assemblages. Consider that two components AM and BM form the crystalline solution (A,B)M and two other components form (A,B)N. These solutions are continuous at high temperatures while at low temperatures they unmix to form coexisting crystalline solutions. These may be regular solutions. The distribution of A and B between the coexisting phases is shown in

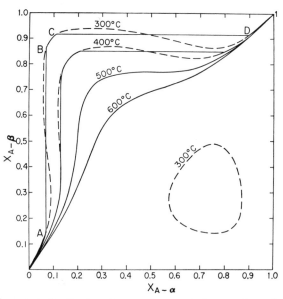

Fig. 7. Theoretical distribution diagram for coexisting regular solutions. Dashed portions of curves indicate metastable and unstable compositions in which immiscibility occurs. Immiscibility gaps are shown as straight full lines. Fig. from MUELLER (1964)

Fig. 7. At $300° C$ as $x_A\left(\dfrac{A}{A+B}\right)$ in $\alpha((A,B)M)$ and $x_A\left(\dfrac{A}{A+B}\right)$ in $\beta((A,B)N)$ change from 0 to 1, we have solution between 0 and A, a miscibility gap between A and B, solution between B and C, a miscibility gap between C and D, and finally solution between D and 1.

Let us consider a qualitative illustration of the behaviour of diopside, enstatite, tremolite and cummingtonite (or anthophyllite) in the system $CaSiO_3-MgSiO_3-SiO_3H_2$ as shown in Fig. 8. The calcium and magnesium components form very non-ideal solutions and miscibility gaps

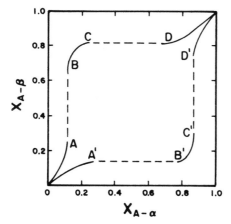

Fig. 8. Qualitative illustration of alternative sequences of coexisting solution based on immiscibility model. (After MUELLER, 1964)

occur in both amphiboles and pyroxenes. Let β in Fig. 8 represent composition of the amphiboles between $Ca_2Mg_5Si_8O_{22}(OH)_2$ and $Mg_7Si_8O_{22}$ $(OH)_2$ and α represent composition of the pyroxenes between $CaMgSi_2O_6$ and $Mg_2Si_2O_6$. Two different types of curves 0–ABCD–1 or 0–A'B'C'D'–1 are possible depending upon whether amphiboles or the pyroxenes develop the miscibility gap first. These two alternative sequences of mineral assemblages are (see MUELLER, 1964):

0–A, tremolite-diopside; A–B, tremolite-diopside-kupfferite (magnesio-anthophyllite); B–C, diopside-kupfferite; C–D, diopside-kupfferite-enstatite; D–1, kupfferite-enstatite.

0–A', tremolite-diopside; A'–B', tremolite-diopside-enstatite; B'–C', tremolite-enstatite; C'–D', tremolite-enstatite-kupfferite; C'–1, enstatite-kupfferite.

Note that in the first sequence tremolite-enstatite and in the second sequence diopside-kupfferite are excluded. The natural assemblage contains minerals, which are not binary solutions and the presence of other ions in significant concentration changes the solution properties of the minerals. The presence of Fe^{2+}, however, does not change the simple scheme mentioned above and the first of the two sequences is well developed in iron formation assemblages described by MUELLER (1960) and KRANCK (1961). The presence of Al^{3+}, however, may cause a change in the scheme as indicated by the stable presence of orthopyroxene-hornblende (corresponding to enstatite-tremolite) assemblages in many rocks of the granulite facies of metamorphism.

b) Intrinsic and Extrinsic Stability

The absence of a mineral from an assemblage may of course be due to other reasons involving intensive variables such as partial pressure of H_2O. The energetics discussed above involve the *intrinsic instability* of the crystalline solution which is the result of the positive excess free energy of mixing G^{EM} associated with the solution. As discussed before, below a critical solution point, G^{EM} so increases that it results in an upwarping of the free energy surface in a certain region of the composition and causes the spinodal decomposition of the solution. It is evident that an ideal crystalline solution is always intrinsically stable, and its absence from an assemblage may be due to the change in the physical-chemical conditions in such a way that certain reaction products form a lower free energy assemblage than the crystalline solution. This instability following MUELLER (1964) may be called *extrinsic*. A solution may be both intrinsically and extrinsically unstable.

Olivine $(Mg, Fe)_2SiO_4$ and pyroxene $(Mg, Fe)SiO_3$ may be considered as ideal binary solutions at high temperatures ($\sim 1100°$ C). In spite of their nearly ideal character at high temperatures, orthopyroxenes with more than 55 mole percent of ferrosilite were found unstable at liquidus temperatures by BOWEN and SCHAIRER (1935). The iron-rich pyroxene is unstable because of the extrinsic instability of ferrosilite relative to fayalite and quartz. At low temperatures ($\sim 600°$ C) orthopyroxene is somewhat non-ideal, and high values of G^{EM} are associated with the high ferrosilite content of the solution. The extrinsic instability of the solution relative to olivine and quartz is less because iron-rich pyroxenes (~ 86 mole percent $FeSiO_3$) are stable in metamorphic rocks. The instability of pyroxenes with higher ferrosilite in metamorphic rocks may be due to both the extrinsic and intrinsic instability of the crystalline solution.

c) Immiscibility in Garnets

Garnets, $(Mg, Fe, Mn, Ca)_3Al_2Si_3O_{12}$, are important rock-forming minerals. They have a large $P - T$ and compositional stability field and therefore occur in a wide variety of metamorphic rocks. The composition of the garnets has been used, either alone or along with the composition of a coexisting mineral, as indicator of $P - T$ of the formation of the host rocks. In such studies, it is often assumed that garnet is an ideal crystalline solution of the pure end members. There are some indications, however, that garnets may not be ideal.

Crystal chemically the ions Fe^{2+} and Mn^{2+}, are significantly different from Ca^{2+} and Mg^{2+}. These ions occupy 8-coordinated sites which are somewhat too large for Mg^{2+} and somewhat too small for Ca^{2+}. The unit cell volume of grossular is considerably larger than the unit cell volumes of the pyralspite series. Besides ionic size, Ca^{2+} is also significantly lower in ionization potential than the other three ions. The above mentioned criteria do not necessarily mean that the solution of species will be non-ideal but they increase the probability of non-ideality.

Another indication of the non-ideality in garnet is provided by the study of the distribution of Mg^{2+} and Fe^{2+} between garnet and a coexisting mineral. Several workers (ALBEE, 1965; KRETZ, 1961; SEN and CHAKRABORTY, 1968; SAXENA, 1968a) find that the distribution coefficient is a function of the concentrations of Mn and Ca in garnet. This would be true even if the quaternary solution was ideal. However it is found that even a small change in the concentration of Mn may change the distribution coefficient significantly, particularly when there is little Mg present in garnet. Such effects are likely to be due to the non-ideality of the Mg–Mn garnets.

Similar to Ca and Mg end members in amphiboles, pyroxenes and olivine, we expect that Mg-garnet and Ca-garnet would form highly non-ideal solutions. No systematic experimental work has been done in determining immiscibility in garnets but according to some preliminary work by GENTILE and ROY (1960) there is no solid solubility between pyrope $(Mg_3Al_2Si_3O_{12})$ and spessartine $(Mn_3Al_2Si_3O_{12})$ or grossular $(Ca_3Al_2Si_3O_{12})$ (temperature not mentioned). Note that in these experiments the whole composition range was not covered. HRICHOVA (1968) reports a partial solubility between spessartine and grossular at $1100°$ C. Therefore it is possible to consider that solvii exist in the ternary and quarternary garnet systems and the reason that coexisting garnets have not been reported is due to the *extrinsic* instability of grossular with respect to reaction products.

Unfortunately in absence of data on actually coexisting garnets, no information can be obtained on the thermodynamic properties of the crystalline solution. It is, however, possible to consider certain hypothetical systems which may be qualitatively of the same nature as the garnet system. Fig. 9 shows the solvus relations in a ternary system $(A, B, C)M$. The constants used in calculating the tie lines and the solvus

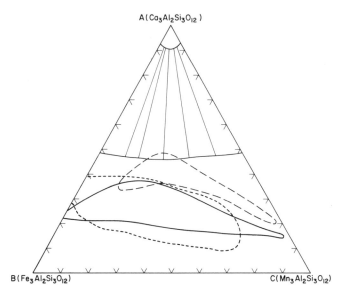

Fig. 9. Miscibility gap in a hypothetical ternary solution $(A, B, C)M$ superimposed on a ternary compositional plot of metamorphic garnets ($\sim 400°$ C) as given by BROWN (1967, 1970). Solid line encloses composition range of garnets from the Sanbagawa belt, Japan. Line with short dashes encloses composition range of garnets from Franciscan, California and the line with long dashes encloses composition ranges of garnets from eastern Otago. For references on localities see BROWN (1967, 1970). Activity-composition relations in garnets are likely to be of the same form as in the hypothetical solution $(A, B, C)M$. The solution parameters are $E_{12} = 0, E_{21} = 3500, E_{13} = 0, E_{31} = 3500, E_{23} = 0, E_{32} = 0, C = 0$ and $T = 673°$K

curves are (see Eqs. (III.31), (III.32) and (III.33) and the computer program Terngap in Appendix):

$$E_{AB} = 0.0, \ E_{BA} = 0.0, \ E_{AC} = 0.0, \ E_{CA} = 3500, \ E_{BC} = 0.0, \ E_{CB} = 3500, \ C = 0.0.$$

Let us assume that AM, BM and CM correspond to $Fe_3Al_2Si_3O_{12}$, $Mn_3Al_2Si_3O_{12}$ and $Ca_3Al_2Si_3O_{12}$ respectively and plot the garnet compositions on this figure. If the crystalline solution $(A, B, C)M$ is similar in thermodynamic properties to the garnet crystalline solution and if the solvus in Fig. 9 represents $400°$ C, none of the garnet compositions from low to medium grade of metamorphism (epidote-amphibolite facies) may lie in the miscibility gap.

Chemical composition of garnets as given by BROWN (1967, 1970) from the Otago schists and from several other areas as quoted by BROWN (1967) are plotted in Fig. 9. The Otago garnets (BROWN, 1970) are selected for the present illustration because of several reasons. First, pyrope ($Mg_3Al_2Si_3O_{12}$) content of the garnets is small and the compositions closely represent the ternary almandine ($Fe_3Al_2Si_3O_{12}$)-spessartine ($Mn_3Al_2Si_3O_{12}$)- grossular ($Ca_3Al_2Si_3O_{12}$) system. Second, the analyses represent the composition of individual spots rather than the average composition of zoned garnets. Third, the metamorphic grade varies only little and the triangular diagram may be regarded as isothermal ($\sim 400°$ C). Fourth, the compositional variation covers a wide range and the garnets may be regarded as fairly good representatives of garnets from widely different localities formed under similar P and T conditions. Fig. 9 shows that the compositions of the garnets do not lie in the region of the miscibility gap for the ternary crystalline solution $(A, B, C)M$ and it may be expected that thermodynamic properties of the ternary garnets are at least qualitatively similar to the properties of the solution $(A, B, C)M$. The binary solution $(A, B)M$ corresponding to (Fe, Mn)-garnet is assumed to be ideal and the constants for the other two binaries $(A, C)M$ and $(B, C)M$ corresponding to (Fe, Ca)-garnet and (Mn-Ca)-garnet respectively are chosen to form coexisting crystalline solutions one of which is close to the extrinsically instable end member phase CM ($\equiv$ grossular). This requires that both $(A, C)M$ and $(B, C)M$ are asymmetric crystalline solutions. Fig. 10 shows the activity-composition relations in $(A, C)M$ or $(B, C)M$ which should be qualitatively of the same form as those of (Fe, Ca)-garnet and (Mn-Ca)-garnet in rocks of epidote-amphibolite facies. In the binary solutions grossular ($Ca_3Al_2Si_3O_{12}$) and almandine ($Fe_3Al_2Si_3O_{12}$) or grossular and spessartine ($Mn_3Al_2Si_3O_{12}$) higher excess free energy G^{EM} is associated

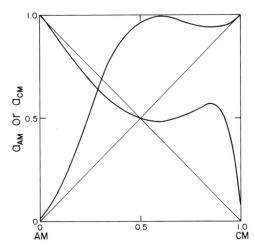

Fig. 10. Activity-composition relation in binary asymmetric solution $(A, C)M$.
$A_0 = 1750, A_1 = 1750$ cal/mole

with high grossular content in the crystalline solution. This energy is evidently required to keep the symmetry and structure of the crystalline solution intact as it is enriched in an ion (Ca^{2+}) larger in size and lower in ionization potential than the ions (Fe^{2+}, Mn^{2+}) occupying the dodecahedral sites.

In absence of any quantitative data on the activity-composition relations in garnet, it is significant to know the form of non-ideality in garnet, even if it is qualitative. From the relations discussed here, it would appear that Ca–Mg-garnets would also be non-ideal asymmetric solutions and this non-ideality would be significant in garnets crystallizing at pressures and temperatures of metamorphism.

IV. Composition of Coexisting Phases

1. Ideal Solution Model

a) Distribution of a Component between Two Ideal Binary Crystalline Solutions

Although there are no strictly binary silicates, certain minerals such as orthopyroxene and olivine may be assumed to be quasi-binary. As Fe^{2+} and Mg^{2+} are similar in ionic charge and size, olivine and orthopyroxene may be assumed to be binary ideal solutions. This assumption will be reexamined later.

RAMBERG and DEVORE (1951) considered the following ion-exchange equilibrium between olivine and pyroxene:

$$MgSiO_3 + 1/2\, Fe_2SiO_4 \rightleftharpoons FeSiO_3 + 1/2\, Mg_2SiO_4. \qquad (IV.1)$$

The equilibrium constant for this reaction at a certain P and T is

$$K_{IV.1} = \frac{x_{Fe\text{-}opx}(1 - x_{Fe\text{-}ol})}{(1 - x_{Fe\text{-}opx})\, x_{Fe\text{-}ol}}. \qquad (IV.2)$$

The equilibrium constant K is a function of P and T only. In the present case, however, $K_{IV.1}$ is not found to be constant except at high temperatures (see OLSEN and BUNCH, 1970).

It may be noted that Eq. (IV.1) is written on a one-cation exchange basis. It may also be written as

$$2\,MgSiO_3 + Fe_2SiO_4 \rightleftharpoons 2\,FeSiO_3 + Mg_2SiO_4. \qquad (IV.3)$$

The equilibrium constant for this reaction is

$$K_{IV.3} = \frac{x_{Fe\text{-}opx}^2 (1 - x_{Fe\text{-}ol})}{(1 - x_{Fe\text{-}opx})^2\, x_{Fe\text{-}ol}}.$$

A Roozeboom figure with such K values has been presented by KERN and WEISBROD (1967, p. 224). It is known empirically from the distribution data in several mineral assemblages that equilibrium constants or distribution coefficients such as $K_{IV.3}$ are very cumbersome to handle and inconsistent with petrological observations. One may, therefore, prefer to use the distribution data on a one-cation exchange basis. It is obvious that in actual calculations of the energy values, it will be necessary to adjust for the activity-composition relations as discussed before in Chapter I.

Generally olivine and pyroxene coexist with several other minerals of fixed or variable composition. If any change in the concentration of the minor components does not change the quasi-binary character of the two minerals, $K_{IV.1}$ is not a function of any changes in the number or proportion or composition of other coexisting phases. This is generally true about equilibrium constants in other systems also. At a certain P and T the stability of the olivine and pyroxene combination is a function of the presence or absence of quartz, but the value of $K_{IV.1}$ itself is not affected.

KRETZ (1959) used ROOZEBOOM plots extensively to show the orderly distribution of cations between coexisting silicate minerals in rocks. If chemical equilibrium is closely approached in the distribution of a component between two binary solutions at a certain P and T, the distribution isotherm is a smooth curve. If at the same time both the solutions are ideal, it will be of the form shown in Fig. 11.

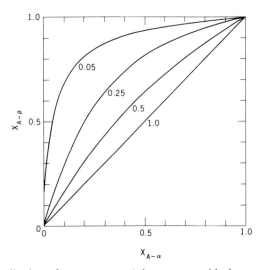

Fig. 11. Distribution of a component A between two ideal crystalline solutions α and β. The numerical values are equilibrium constants

b) Coexisting Ternary Ideal Solutions

Consider two coexisting ternary phases α and β with the formulae $(A, B, C)M$ and $(A, B, C)N$. The pure components are AM, BM, and CM in α and AN, BN, and CN in β. The chemical potentials of the components in α and β are

$$\mu_{AM-\alpha} = \mu^0_{AM-\alpha} + RT \ln x_{AM-\alpha} \tag{IV.4}$$

and

$$\mu_{AN-\beta} = \mu^0_{AN-\beta} + RT \ln x_{AN-\beta} . \tag{IV.5}$$

The other potentials are similarly defined. The potentials of all the pure components μ are functions of P and T only. $x_{A-\alpha}$ and $x_{A-\beta}$ may be substituted for $x_{AM-\alpha}$ and $x_{AM-\beta}$, respectively, without altering the results (see discussion before).

The distribution of A between α and β may be represented by the ion exchanges

$$A\text{-}\alpha + B\text{-}\beta \rightleftharpoons B\text{-}\alpha + A\text{-}\beta \tag{IV.6}$$

and

$$A\text{-}\alpha + C\text{-}\beta \rightleftharpoons C\text{-}\alpha + A\text{-}\beta . \tag{IV.7}$$

The equilibrium constants may be written as

$$K_{IV.6} = \frac{x_{B-\alpha} x_{A-\beta}}{x_{A-\alpha} x_{B-\beta}}$$

and

$$K_{IV.7} = \frac{x_{C-\alpha} x_{A-\beta}}{x_{A-\alpha} x_{C-\beta}}$$

where $x_A = A/(A + B + C)$, and the other x's are defined similarly. Both $K_{IV.6}$ and $K_{IV.7}$ will be constants for all ratios of A to B to C. A plot of $x_{A-\alpha}$ against $x_{A-\beta}$ will produce a symmetric ideal distribution curve.

2. Nonideal Solutions

a) Distribution of a Component between Two Simple Mixtures

For the ion-exchange equation

$$A\text{-}\alpha + B\text{-}\beta \rightleftharpoons A\text{-}\beta + B\text{-}\alpha \tag{IV.6}$$

at equilibrium,

$$\mu_{B-\beta} + \mu_{A-\alpha} = \mu_{B-\alpha} + \mu_{A-\beta} . \tag{IV.8}$$

If (A, B) M and (A, B) N are simple mixtures,

$$\mu_{B-\beta} = \mu^0_{B-\beta} + RT \ln(1 - x_{A-\beta}) + W_\beta (x_{A-\beta})^2 \tag{IV.9}$$

and the other μ's are similarly defined. Substituting the values found by Eq. (IV.9) into Eq. (IV.8) and rearranging,

$$\ln \frac{x_{A-\beta}(1 - x_{A-\alpha})}{(1 - x_{A-\beta}) x_{A-\alpha}} - \left[\frac{W_\alpha}{RT}(1 - 2x_{A-\alpha}) - \frac{W_\beta}{RT}(1 - 2x_{A-\beta}) \right] = \frac{-\Delta G^0_{IV.6}}{RT}$$

$$\tag{IV.10}$$

where

$$-\Delta G^0_{IV.6} = \mu^0_{B\text{-}\beta} + \mu^0_{A\text{-}\alpha} - \mu^0_{B\text{-}\alpha} - \mu^0_{A\text{-}\beta}.$$

Or

$$\ln K_{IV.6} = \ln K_D - \frac{W_\alpha}{RT}(1 - 2x_{A\text{-}\alpha}) + \frac{W_\beta}{RT}(1 - 2x_{A\text{-}\beta}) \qquad (IV.11)$$

where $K_{IV.6} = \exp(-\Delta G^0_{IV.6}/RT)$ and K_D is the distribution coefficient.

If a good least-squares fit can be obtained for the distribution data by using Eq.(IV.11), it may be found that both minerals are close to being simple mixtures.

As one or both of the minerals become less ideal, the distribution isotherms may attain different forms (see MUELLER, 1964). Fig. 12 shows an example where one of the minerals α is ideal and β is nonideal. W_β is assumed to vary linearly with $1/T$. The values of W_β/RT and $K_{IV.6}$ at 673 and 1673 °K are 2.75, 1.603, 0.77, and 1.518, respectively (see SAXENA, 1969a). The forms of the distribution isotherms are very different from the symmetric ideal curves.

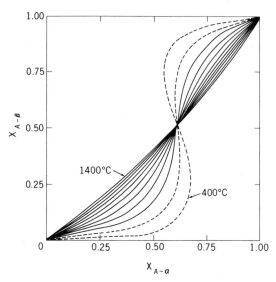

Fig. 12. Distribution of a component A between an ideal solution α and a regular solution β. The data are

T °K	$\dfrac{W^\beta}{RT}$	K
673	2.75	1.60
1173	0.77	1.52

b) Coexisting Regular Ternary Solutions

The composition of two coexisting phases that obey the same equation of state is considered here as an example. These phases are products of unmixing in a ternary solution $(A, B, C)M$. W'_{AB}, W'_{BC}, and W'_{AC} are assumed to be 1500, 7000, and 9000 cal/mole respectively, and the values of W' are assumed not to be functions of P, T, and composition (regular solution). Fig. 13 shows the miscibility gap in the system and the tie

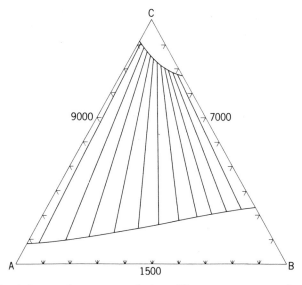

Fig. 13. Coexisting regular ternary solutions. The components are A, B, and C. W'_{AB}, W'_{BC} and W'_{AC} are 1500, 7000 and 9000 cal/mole respectively. The temperature is assumed to be $1573\,°K$

lines for the coexisting phases. Let the phase rich in C be denoted by α and the phase poor in C by β. For the chemical potentials,

$$\mu_{A\text{-}\alpha} = \mu^0_{A\text{-}AM} + RT \ln f_{A\text{-}\alpha} x_{A\text{-}\alpha} \qquad (IV.12)$$

and

$$\mu_{A\text{-}\beta} = \mu^0_{A\text{-}AM} + RT \ln f_{A\text{-}\beta} x_{A\text{-}\beta}. \qquad (IV.13)$$

Any one of the following ion exchanges between α and β may be considered:

$$A\text{-}\beta + B\text{-}\alpha \rightleftharpoons A\text{-}\alpha + B\text{-}\beta,$$

$$B\text{-}\alpha + C\text{-}\beta \rightleftharpoons B\text{-}\beta + C\text{-}\alpha, \qquad (IV.14)$$

and

$$A\text{-}\alpha + C\text{-}\beta \rightleftharpoons C\text{-}\alpha + A\text{-}\beta.$$

Table 1. Composition of coexisting phases in ternary regular solutions

B	A	C	B	A	C	$x_{B-\alpha}$	$x_{B-\beta}$	K_D
0.010	0.082	0.908	0.043	0.868	0.089	0.108	0.047	0.406
0.040	0.079	0.881	0.174	0.718	0.109	0.336	0.195	0.478
0.080	0.071	0.849	0.339	0.522	0.140	0.530	0.394	0.576
0.100	0.066	0.834	0.413	0.431	0.156	0.602	0.489	0.632
0.130	0.056	0.814	0.512	0.309	0.179	0.699	0.624	0.715
0.170	0.037	0.793	0.625	0.171	0.205	0.821	0.785	0.796
0.200	0.020	0.780	0.700	0.081	0.219	0.909	0.896	0.862

$$x_B = \frac{B}{B+A} \quad \text{and} \quad K_D = \frac{x_{B-\alpha} x_{A-\beta}}{x_{B-\beta} x_{A-\alpha}}.$$

The equilibrium constant for first of the Eq. (IV.14) is

$$\frac{x_{A-\alpha} x_{B-\beta}}{x_{A-\beta} x_{B-\alpha}} \cdot \frac{f_{A-\alpha} f_{B-\beta}}{f_{A-\beta} f_{B-\alpha}} = K_{IV.14} \qquad (IV.15)$$

In this particular case because α and β obey the same equation of state, $\Delta G^0_{IV.14} = 0$ and $K_{IV.14} = 1$. In other cases where α and β are minerals with different crystal structures, the equilibrium constant is not necessarily

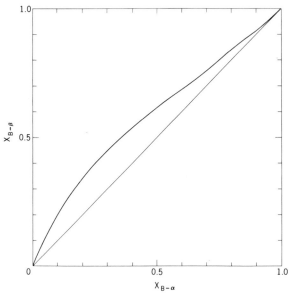

Fig. 14. Distribution of a component B between two ternary regular solutions plotted on a Roozeboom diagram. x is $B/(A+B)$

equal to 1. The f terms in Eq. (IV.15) are functions of P, T, and the ratio of A to B to C. Therefore $K_D(x_{B-\beta} x_{A-\alpha}/x_{A-\beta} x_{B-\alpha})$ also changes with P, T, and the ratios of A to B, B to C, and A to C.

Let the ratio of A to B to C change systematically as listed in Table 1. A plot of $x_{B-\alpha}$ against $x_{B-\beta}$ where x is either A/(A + B) or A/(A + B + C), shows a smooth distribution curve (Fig. 14). The form of the curve, however, is markedly different from the ideal distribution curve.

The activity coefficients are given by

$$RT \ln f_A = (x_B)^2\, W'_{AB} + (x_C)^2\, W'_{AC} + x_B x_C\, (W'_{AB} - W'_{BC} + W'_{AC}), \qquad \text{(IV.16 a)}$$

$$RT \ln f_B = (x_C)^2\, W'_{BC} + (x_A)^2\, W'_{AB} + x_C x_A (W'_{BC} - W'_{AC} + W'_{AB}), \qquad \text{(IV.16 b)}$$

$$RT \ln f_C = (x_A)^2\, W'_{AC} + (x_B)^2\, W'_{BC} + x_A x_B (W'_{AC} - W'_{AB} + W'_{BC}). \qquad \text{(IV.16 c)}$$

where $x_A = A/(A + B + C)$ and the other x's are similarly defined. It may be checked that substitution of f values into Eq. (IV.15) gives the equilibrium constant as unity.

3. Distribution of a Cation between Two or More Multicomponent Minerals

Many rock-forming minerals are complex multicomponent crystalline solutions. The distribution of cations in two or more coexisting minerals in natural assemblages may still yield certain valuable information. The method to be followed in such cases has been discussed by KRETZ (1959). In silicates there are at least two types of coordination for cations. Si^{4+}, Al^{3+}, Fe^{3+}, and less commonly Ti^{4+} are in tetrahedral coordination. Fe^{2+}, Mg^{2+}, Fe^{3+}, Al^{3+}, Mn^{2+}, and Ti^{4+} are found in octahedral coordination. Such differently coordinated ions may be regarded as forming submixtures. The distribution of Fe^{2+} or Mg^{2+} or any other octahedrally coordinated ion may be examined in two or more such submixtures forming parts of different minerals. It should be noted, however, that the chemical potentials of a cation in octahedral coordination may also be a function of any chemical variation in the concentrations of the tetrahedrally coordinated ions. Such information can be usually obtained beforehand by considering the chemical composition of individual minerals. For example, the positive correlation between the concentration of tetrahedrally coordinated Al^{3+} in amphiboles and biotite with the Fe^{2+}/Mg^{2+} ratio in the mineral is now well known (RAMBERG, 1952 b; SAXENA, 1968 a).

It may be argued that the study of the distribution of a component between only two of the coexisting minerals that are quasi-binary solutions out of an entire assemblage of five or six minerals could not be useful. That is, the presence or absence of a third or fourth mineral in the assemblage ought to affect the distribution coefficient. This is not generally true. The distribution coefficient changes only when the presence or absence of a third mineral is associated with a significant change in the concentration of one or more elements in one or both of the coexisting minerals. For example, TiO_2 is only sparingly soluble in olivine and orthopyroxene. The chemical potential of TiO_2 may increase or decrease in the rock, and rutile may be added or removed from the assemblage, but K_D for the distribution of Fe^{2+} and Mg^{2+} would not change. However, if the change in μ_{TiO_2} changes the concentration of TiO_2 significantly in one of the two coexisting minerals, K_D may also change. Thus it is only meaningful to consider the concentrations of all the compo-

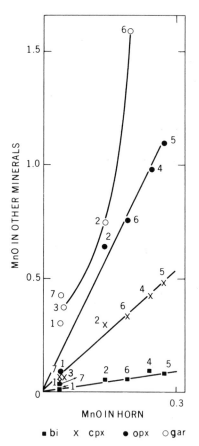

Fig. 15. Distribution of Mn in coexisting minerals in charnockites of Varberg, Sweden (SAXENA, 1968b)

nents in the two minerals and not the presence or absence of another phase or the change in the bulk composition of the rock.

One of the important results of the study of cation partitioning is the recognition of how closely chemical equilibrium may be approached in the rocks. Whether the minerals are ideal or not, the distribution of a component between two coexisting binary phases at a certain P and T will be represented by a smooth distribution curve if chemical equilibrium is closely approached. If the minerals are not binary, the concentration of other components because of the diadochic or substitutional relationships may affect the orderly distribution as discussed before. In fact, the approach to chemical equilibrium can be studied with respect to each component individually. Fig. 15 shows the distribution of Mn in coexisting minerals from charnockites (SAXENA, 1968 b). Such orderly distribution of Mn is common in other rocks as well. The distribution of Fe^{2+} and Mg^{2+} between coexisting olivine and orthopyroxene at 1073 and 1173 °K was experimentally studied by MEDARIS (1969). Although MEDARIS made repeated grinding and heating of the reaction products, Fig. 16 shows that the distribution points both at 1073 and 1173 °K show some scatter. The difficulties are related to the kinetics of

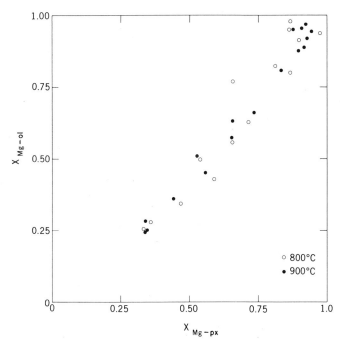

Fig. 16. Distribution of Fe^{2+} and Mg^{2+} between synthetic olivine and pyroxene. Distribution data from MEDARIS (1969)

the ion-exchange reaction as equilibrium is approached, particularly when the distribution approaches a 1 to 1 ratio in the two minerals. In contrast to these experimental results, the partitioning of Mg^{2+} and Fe^{2+} between orthopyroxene and Ca pyroxene (Capx) in metamorphic rocks as studied by KRETZ (1963) is remarkably orderly. Most distribution points fall on a smooth curve (see Fig. 63 later) and a distribution curve representing igneous rocks is clearly separated from a distribution curve for the metamorphic rocks.

Studies of partitioning of cations between coexisting minerals in natural rocks by petrologists (ALBEE, 1965; ANNERSTEN, 1968; BINNS, 1962; BUTLER, 1969; KRETZ, 1959; GORBATSCHEV, 1969; HIETANEN, 1971; and MUELLER, 1960, among many others) are attempts to rationalize the concept of metamorphic facies on a mineral and chemical basis. In experimental systems, similar attempts have been made by NAFZIGER and MUAN (1967), WILLIAMS (1971), LARIMER (1968), and MEDARIS (1969), among others. The results of such partitioning studies have generally confirmed the usefulness of the approach and the need for more thermodynamic data on crystalline solutions.

In essence, problems of phase equilibria are distribution problems, and a statistical approach to such problems may be made to avoid the consideration of the thermodynamic properties of solutions in individual minerals. Such approaches have been made principally by GREENWOOD (1967) and PERRY (1973) and should be applicable in solving the petrogenetic problem of incompatible assemblages and the recognition of chemical equilibrium in natural or experimental systems.

4. Composition of Coexisting Minerals and Chemical Equilibrium in Igneous Rocks

In any rock whether metamorphic or igneous, chemical equilibrium is generally recognized by phase rule considerations. However, when crystalline solutions are involved, we have to consider not only the phase rule, but also equilibrium in the distribution of elements in the coexisting phases. As indicated before, the chemical data on the distribution of elements between coexisting phases in metamorphic rocks assembled in the last decade bear ample testimony to the fact that chemical equilibrium is closely approached in large volumes of the high grade metamorphic rocks. These thermodynamic principles should be equally important in the study of igneous rocks. But there are several important differences in the processes taking place in the igneous and in the metamorphic rocks. The most important is the fact that in metamorphic rocks we are mostly concerned with solid state diffusion whereas in igneous rocks most reactions take place through the liquid medium. As such, most crystals are not in the place where they were nucleated and grown partly or wholly. Thus in igneous rocks, it is not only the ions, as in metamorphic rocks, which move about but the minerals themselves are free to mix with crystals formed earlier or later in the time sequence. The time sequence of crystallization is important because of the continuously changing temperature. This problem, however, is not serious in many plutonic rocks, such as in the layered intrusions in Skaergaard or Stillwater because for the purpose of studying chemical equilibrium, each layer may be considered individually. It is then not unlikely that within each layer the distribution of elements in the phases in contact is orderly indicating that chemical equilibrium was attained.

It does not concern us whether the chemical equilibrium in the distribution of a cation as shown by the composition of the coexisting phases was achieved by a direct ion exchange in the solid state or whether it took place via an intercumulus liquid. In either case it will be formally correct to write an ion exchange type reaction between crystalline solutions. We shall have to consider two types of ion exchange reactions. First is the heterogeneous reaction that takes place between coexisting phases whether crystalline, liquid or gaseous. At any one time such reactions proceed towards minimizing the free energy of the system by redistributing the species among the coexisting phases. The second is the homogeneous reaction that takes place within the crystalline or liquid solutions. These reactions may be termed interphase- and intraphase-ion exchange reactions respectively. The study of both these reactions yield important information on the petrogenesis of igneous rocks. The interphase exchange reactions have been described before. The intraphase reaction involves the exchange of cations among the nonequivalent structural sites in a crystal as a function of temperature and composition and is considered later.

The fact that coexisting phases in igneous rocks may show an equilibrium distribution of cations was noted by KRETZ (1961). Fig. 65 shows that the distribution of Fe^{2+}–Mg is orderly in coexisting igneous pyroxenes and is consistent with the findings in metamorphic rocks. Each distribution point represents an average composition of several pyroxene grains separated from a rock specimen. It was, therefore, indeed remarkable that the distribution curve is smooth for samples from different igneous rocks occurring in widely separated geographic locations. McCALLUM (1968) studied the rocks of the Stillwater Complex. In such layered intrusions it is commonly found that the time sequence of crystallization can be distinguished into two stages — 1) cumulus stage during which one or more minerals crystallized from the magma and settled to the floor

2) the postcumulus stage during which the cumulus minerals were modified by overgrowths or reactions with the intercumulus liquid accompanied by direct precipitation of other minerals from this liquid. A third category of reactions may take place in the subliquidus-subsolidus stage during which the solid phases partially reequilibrate in response to the falling temperature.

Petrographic and textural characters indicate that in a particular horizon at any one time the cumulus stage and the postcumulus stage are largely mutually exclusive. This is indeed very significant since if we can consider a small volume in a particular layering, we should be certain

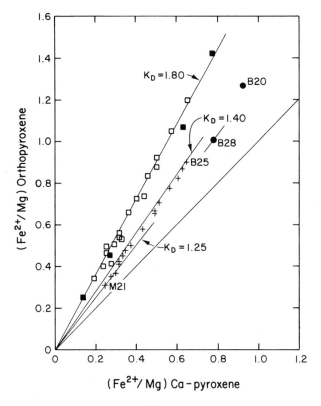

Fig. 17. Relation between the Fe^{2+}/Mg ratios (weight percent) in coexisting pyroxenes from the Stillwater complex. The 45° line is shown for reference. + coexisting orthopyroxene and Ca-pyroxene macrocrystals. ● coexisting "inverted pigeonite" and Ca-pyroxene macrocrystals. □ adjacent host—lamellae pairs with Ca-pyroxene host and orthopyroxene lamellae. ■ adjacent host—lamellae pairs with orthopyroxene or "inverted" pigeonite host and Ca-pyroxene lamellae.
Fig. from McCallum (1968)

that the minerals under chemical study do coexist both in space and time of crystallization.

Fig. 17 shows the distribution of Fe^{2+} and Mg in coexisting pyroxenes from the Stillwater Complex. Distribution points represent coexisting opx and cpx macrocrystals ($+$), 'inverted pigeonite' and cpx macrocrystals ($\bullet$), adjacent host-lamellae pairs — cpx host and opx lamellae ($\square$) opx host and cpx lamellae ($\blacksquare$). These distributions yield information on several aspects of petrogenesis. Note that the macrocrystals (cumulus or early postcumulus) are in chemical equilibrium. The scatter noted here is mainly due to temperature variation and MCCALLUM, assuming pyroxenes to be ideal, estimated a temperature-range between 1250—1010° C. Similarly it may be noted that the distribution between lamallae and host indicate a subsolidus intraphase equilibration with falling temperatures and these distributions may represent temperatures between 600—800° C.

5. Physico-Chemical Conditions of Formation of Mineral Assemblages as Inferred from Composition of Coexisting Phases

The composition of coexisting phases may in suitable cases be used to obtain important petrogenetic information on the P, T and chemical conditions of crystallization of the mineral assemblages. The following discussion is based on the works of BUTLER (1969), BONNICHSEN (1970), KRETZ (1964) and MUELLER (1964, 1966).

The Gibbs phase rule may be stated as:

$$\text{variance} = n + 2 - p$$

where variance is the number of intensive variables that can be changed independently without either the disappearance of a phase or the appearance of a new one, p is the number of phases, and n is the number of independently variable components of the assemblage. Consider the assemblage quartz (SiO_2), orthopyroxene ($FeSiO_3$–$MgSiO_3$), Ca-pyroxene ($CaSiO_3$–$MgSiO_3$–$FeSiO_3$) and calcite ($CaCO_3$). The number of phases present in the assemblage is 4, although there must have been a gas or "intergranular" phase present when equilibrium was established in the system. The number of components is 5. This gives us a variance of 3 and at a given P, T, and μ_{CO_2} the mole fractions of hedenbergite ($CaFeSi_2O_6$) and ferrosilite ($FeSiO_3$) in pyroxenes would be inversely related.

Enumeration of the phases and components may not be simple. In order to avoid the difficulty, KRETZ (1964) used the following definition of variance:

number of independent intensive variables which must be specified to describe the state of the system (i.e. temperature, pressure, mole fractions necessary to define the composition of all crystalline solutions, and number of chemical potentials which are variable when the system is not closed)	− number of equations which must be specified to satisfy the requirements of thermodynamic equilibrium (i.e. equations of exchange equilibrium, equations of phase transformation and other phase equilibria)	= variance

or in the abbreviated form:

number of variables — number of equations = variance.

In the example of the assemblage considered above, we may express the equilibrium as below:

$$FeSiO_3 + CaCO_3 + SiO_2 \rightleftharpoons CaFeSi_2O_6 + CO_2$$

$$MgSiO_3 + CaCO_3 + SiO_2 \rightleftharpoons CaMgSi_2O_6 + CO_2$$

The variance may be calculated as below:

$$T, P, \mu_{CO_2}, x_{\text{Fe-opx}}, x_{\text{Fe-cpx}} \quad - \quad \begin{matrix} G_{CaFeSi_2O_6} + \mu_{CO_2} - G_{FeSiO_3} - G_{CaCO_3} - G_{SiO_2} \\ G_{CaMgSi_2O_6} + \mu_{CO_2} - G_{MgSiO_3} - G_{CaCO_3} - G_{SiO_2} \end{matrix} = 3$$

$$\text{variables} = 5 \qquad \text{equations} = 2$$

Consider the following expressions as presented by BUTLER (1969) for assemblages which in suitable cases may be used to deduce the relationship of μ_{CO_2} and μ_{H_2O} with the variations of mineral composition:

$$H_2O = \underset{\text{gru}}{Fe_7Si_8O_{22}(OH)_2} - 7\,\underset{\text{opx}}{FeSiO_3} - \underset{\text{qtz}}{SiO_2} \tag{a}$$

$$CO_2 = \underset{\text{cal}}{CaCO_3} - \underset{\text{cpx}}{CaFeSi_2O_6} + \underset{\text{opx}}{FeSiO_3} + \underset{\text{qtz}}{SiO_2}, \tag{b}$$

$$H_2O = \underset{\text{act}}{Ca_2Fe_5Si_8O_{22}(OH)_2} - 2\,\underset{\text{cpx}}{CaFeSi_2O_6} - 3\,\underset{\text{mag}}{Fe_3O_4} \tag{c}$$

$$CO_2 = \underset{\text{cal}}{CaCO_3} - \underset{\text{cpx}}{CaFeSi_2O_6} + \underset{\text{mag}}{Fe_3O_4} - \underset{\text{hem}}{Fe_2O_3} + 2\,\underset{\text{qtz}}{SiO_2}. \tag{d}$$

In all the above expressions, the intensive variables are 5 (T, P, μ_{H_2O} or μ_{CO_2}, and two mole fractions x_{Fe}'s) and the number of equations is two, one for Fe end members and one for Mg. This gives us a variance of 3. Therefore the variation in composition of the crystalline solutions in such assemblages within a small area, should be related to the variation in the chemical potentials of CO_2 and H_2O. Note that the small variation in the concentration of Ca in clinopyroxene, actinolite and calcite has been neglected in the above analysis.

It is evident that if we have thermochemical data on the various phases and on the properties of the crystalline solutions, a quantitative determination of the intensive variables is possible. MUELLER (1966) has discussed the physico-chemical conditions of formation of mineral assemblages in iron formations using the following reactions:

$$MgSiO_3 + CaCO_3 + SiO_2 \rightleftharpoons CaMgSi_2O_6 + CO_2 , \qquad \text{(e)}$$
$$_{\text{opx}} \quad {}_{\text{cal}} \quad {}_{\text{qtz}} \quad\quad\quad {}_{\text{cpx}} \quad\quad {}_{\text{fluid}}$$

$$MgSiO_3 + CaFeSi_2O_6 \rightleftharpoons FeSiO_3 + CaMgSi_2O_6 , \qquad \text{(f)}$$
$$_{\text{opx}} \quad\quad {}_{\text{cpx}} \quad\quad\quad {}_{\text{opx}} \quad\quad {}_{\text{cpx}}$$

$$CO_2 \rightleftharpoons C + O_2 . \qquad \text{(g)}$$
$$_{\text{fluid}} \quad {}_{\text{graphite fluid}}$$

If it is assumed that the crystalline solutions opx, and cpx are ideal, the ΔV_f is negligible and calcite is a pure phase, we may write the following equations:

$$K_e \exp\left(\frac{-P\Delta V_e}{RT}\right) = \frac{x_{\text{Mg-cpx}}}{x_{\text{Mg-opx}}} P^*_{CO_2} \qquad \text{(IV.17)}$$

$$K_f = \frac{x_{\text{Mg-cpx}}(1 - x_{\text{Mg-opx}})}{x_{\text{Mg-opx}}(1 - x_{\text{Mg-cpx}})} \qquad \text{(IV.18)}$$

where $P^*_{CO_2}$ refers to the fugacity of CO_2. By elimination of $x_{\text{Mg-cpx}}$ from the Eqs. (17) and (18), we have

$$P^*_{CO_2} = \frac{K_e}{K_f} \exp\left(\frac{-P\Delta V_e}{RT}\right)[(K_f - 1) x_{\text{Mg-opx}} + 1] \qquad \text{(IV.19)}$$

which shows as before that the fugacity of CO_2 is a linear function of the composition of orthopyroxene at constant temperature and total pressure. Actual computation of K_e as a function of T is difficult because of the simplifications introduced above and because the thermochemical data on all phases (orthopyroxene) are not available and finally because there are large uncertainties in the thermochemical data which are available (see MUELLER, 1966, for an approximate $\log K_e$ and T^{-1} relationship).

6. Stability of Orthopyroxene

An example where the change in the composition of a crystalline solution changes the stability relations in the $P - T$ field may now be considered. In metamorphic rocks of the granulite facies orthopyroxene, $(Mg, Fe)SiO_3$, is a stable mineral and the mole fraction of Fe may be as much as 0.86 (specimen XYZ from GREENLAND, RAMBERG, and DeVORE, 1951). In general, however, pyroxenes with $x_{Fe\text{-}opx}$ greater than 0.5 are not common in natural assemblages and this was explained as due to the extrinsic instability of ferrosilite relative to the decomposition products quartz and fayalite. From the experimental work (LINDSLEY, MACGREGOR and DAVIS, 1964; AKIMOTO, KATURA, SYONO, FUJISAWA and KOMADA, 1965) it is established that pressure stabilizes orthoferrosilite. The pressures involved in these experiments are up to about 10 Kbars which is not different from the range of granulite facies conditions of metamorphism (4 to 8 Kbar). It may, therefore, be worth examining the conditions which may increase the stability field of pyroxenes in natural assemblages, for example the effect of adding $MgSiO_3$ to the pyroxene crystalline solution as considered by OLSEN and MUELLER (1966) and KUREPIN (1970). The following two reactions must be considered:

$$Fe_2SiO_4 + SiO_2 = 2\,FeSiO_3, \qquad \text{(h)}$$
$$_{\text{ol}} \quad _{\text{qtz}} \quad _{\text{opx}}$$

$$1/2\,Mg_2SiO_4 + FeSiO_3 = 1/2\,Fe_2SiO_4 + MgSiO_3. \qquad \text{(i)}$$
$$_{\text{ol}} \quad _{\text{opx}} \quad _{\text{ol}} \quad _{\text{opx}}$$

At equilibrium, we have

$$K_h\left(\frac{f_{Fe\text{-ol}}}{f_{Fe\text{-px}}}\right)^2 \exp\left(\frac{-P\Delta V_h}{RT}\right) = \left(\frac{x_{Fe\text{-px}}}{x_{Fe\text{-ol}}}\right)^2, \qquad \text{(IV.20)}$$

$$K_i\left(\frac{f_{Mg\text{-ol}}\,f_{Fe\text{-px}}}{f_{Mg\text{-px}}\,f_{Fe\text{-ol}}}\right)\exp\left(\frac{-P\Delta V_i}{RT}\right) = \frac{(1 - x_{Fe\text{-px}})\,x_{Fe\text{-ol}}}{(1 - x_{Fe\text{-ol}})\,x_{Fe\text{-px}}}, \quad \text{(IV.21)}$$

where ΔV is the volume change for the reaction (h) or (i). ΔV_i is small and uncertain. From Eqs.(IV.20) and (IV.21), the following expression may be derived by eliminating $x_{Fe\text{-ol}}$ as done by OLSEN and MUELLER (1966):

$$P = \frac{189\,T}{(-\Delta V_h)}\{\log[(K_i'\varphi_i^{-1} - 1)x_{Fe\text{-px}} + 1]^2 - \log[K_h\varphi_h^{-2}(K_i')^2\varphi_i^{-2}]\} \quad \text{(IV.22)}$$

where K_i' is $K_i\exp(-P\Delta V_i/RT)$, φ_i and φ_h represent the product and ratios of the activity coefficients in Eqs.(IV.21) and (IV.20) respectively. To know the relation between P (atmospheres) and T from Eq. (IV.22) at a given temperature, we need data on ΔV_h, K_i', K_h and on the solution properties of olivine and orthopyroxene. Uncertainties in one or more of these data may make the calculations merely an exercise. Assuming that both pyroxene and olivine at high temperature are ideal solutions, we may express (IV.22) as

$$P = \frac{0.192\,T}{-\Delta V_h}\ln\left\{\frac{1 - x_{\text{Fe-px}} + K_i' x_{\text{Fe-px}}}{K_i' K_h}\right\} \qquad (IV.23)$$

where P is in Kbar and Eq.(h) is considered on one ion basis. The following data may now be used as was done by KUREPIN (1970):

$$\Delta V_h(\alpha\text{-qtz}) = -1.41\ \text{cm}^3/\text{mole}$$

or

$$\Delta V_h(\beta\text{-qtz}) = -1.50\ \text{cm}^3/\text{mole}$$
$$\Delta V_i = -0.30\ \text{cm}^3/\text{mole}$$

$$K_i' = K_i \exp\left(\frac{-\Delta V_i P}{RT}\right)$$

where K_i is the ideal distribution coefficient or equilibrium constant calculated from BOWEN and SCHAIRER's (1935) and NAFZIGER and MUAN's (1967) data on coexisting olivine and pyroxene at 1100, 1200 and 1305° C;

$$K_h = \exp\left(\frac{-\Delta G_h}{RT}\right),$$

which may be calculated from NAFZIGER and MUAN (1967).

In order to match the equilibrium condition for the reaction (h) as found in the experimental work (LINDSLEY et al., 1964 or LINDSLEY, 1965), it was found necessary to change ΔV_h so that the P calculated from $-\Delta G_h/\Delta V$ is the same as the P in the experimental work. KUREPIN (1970) finally uses $\Delta V_h = -1.3\ \text{cm}^3/\text{mole}$. The $P-T$ and $x_{\text{Fe-px}}$ relationship using the data discussed above is shown in Fig. 18 which may be used as a geobarometer.

It must, however, be remembered that some of the data and assumptions used above may not be valid and there may be other type of relationship in P, T and $x_{\text{Fe-opx}}$. Consider, for example, the assumption of ideal solution in olivine and pyroxene. NAFZIGER and MUAN's (1967) data on the activity-composition relation has been discussed in other chapters. Olivine appears to be somewhat non-ideal at 1200° C. Pyroxenes may show negative excess free energy of mixing at 1000° C which may be indicated from VIRGO and HAFNER's (1969) intra-crystalline ion exchange data. It is interesting to note that in this regard

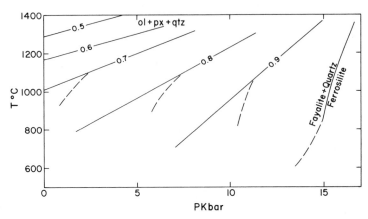

Fig. 18. Stability of orthopyroxene as a function of P, T and $x_{Fe\text{-}opx}$ (modified from KUREPIN, 1970). The lines are from LINDSLEY's (1965) data on fayalite + quartz → ferrosilite with numbers representing $x_{Fe\text{-}opx}$. The calculations are based on the assumption of ideal solution model for both olivine and pyroxene. The dashed curves indicate the probable path if non-ideality in orthopyroxene at lower temperatures ($< 800°$ C) is considered

OLSEN and MUELLER's (1966) calculations show that a negative deviation from ideality in orthopyroxene would greatly enhance the effect given by Eq. (IV.22). At any rate in view of the positive deviation of orthopyroxene from ideality between 600—800° C (see Chapter VIII), extrapolation of the curves in Fig. 18 may not be correct in form. These curves may bend away from the vertical axis as the temperature decreases as shown in the figure.

7. Distribution of Trace Elements

The thermodynamics of the distribution of trace elements between coexisting phases has been considered by MCINTIRE (1963), BURNS and FYFE (1966) and by WHITTAKER (1967). In this section data on the distribution of trace elements are examined from the points of view of chemical equilibrium and of the possible temperature dependence of the distribution.

From the data on the distributions in coexisting minerals in charnockitic rocks (LEELANANDAM, 1970), it appears that chemical equilibrium is closely approached in many cases. Ni and Cr partitioning shows the closest approach to equilibrium. Distribution of V is less orderly and Co is scattered. Co on the other hand, is very orderly distributed between amphibole and biotite in some Swedish rocks

(ANNERSTEN and EKSTRÖM, 1971). Distribution of V between horn-blende and biotite in amphibolites was found to be orderly by KRETZ (1959). Unfortunately the accuracy of the concentrations of the trace components in many cases may not be good and the distribution pattern such as that of V, found by KRETZ (1959), are exceptions rather than the rule. With reasonably accurate data, it is expected that if chemical equilibrium is closely approached in the distribution of elements which are in crystalline solution, a certain elongate cluster of the distribution points should result (see ALBEE, 1965).

McCALLUM (1968) studied the distribution of Ni between coexisting minerals in the rocks from the Stillwater Complex. The distribution between coexisting macrocrystals of clinopyroxene and orthopyroxene is orderly as shown in Fig. 19. The samples represent rocks from the base of the complex to the uppermost exposed horizons. The equilibration temperature decreases from the base to the top. This temperature difference ($\approx 200°$ C) is not noted in the partitioning but it does show that equilibrium was closely approached in the distribution of Ni between the two minerals.

The temperature dependence of the distribution of trace elements is predicted from thermodynamic arguments. However this aspect has not been studied in detail either in experimental or natural silicate mineral assemblages. In this context the distributions of Ni, V, Ga and Ba

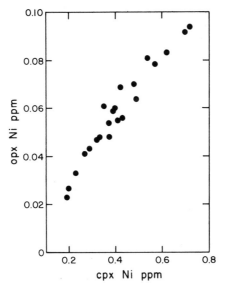

Fig. 19. Distribution of Ni between coexisting orthopyroxene and clinopyroxene from Stillwater igneous complex. (Data from McCALLUM, 1968)

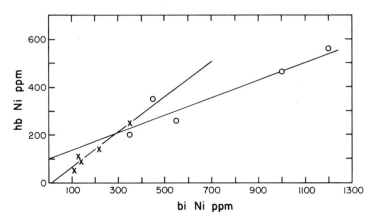

Fig. 20. Distribution of Ni between coexisting hornblende and biotite. Open circles — charnockites (LEELANANDAM, 1971), crosses — igneous plutonic rocks (HIETANEN, 1971)

between coexisting biotite and hornblende in rocks of different meta-morphic grade as shown in Figs. 20—23 are significant. Fig. 20 shows distribution of Ni between the two minerals in charnockites (LEELANAN-DAM, 1970) and in Sierra Nevada plutons (HIETANEN, 1971). Similar distributions for Ba, Ga and V are shown in Figs. 21, 22, and 23 respec-tively. In Fig. 23 the distribution of V between biotite and hornblende in high amphibolite rocks (KRETZ, 1959) is also shown. These distributions show that the igneous samples are generally separated from the meta-morphic samples. The slopes of the least squares fitted curves change significantly from one group of samples to the other. Assuming that this is not an influence of the difference in the concentration of other

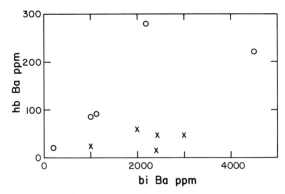

Fig. 21. Distribution of Ba between biotite and hornblende. Samples as in Fig. 20

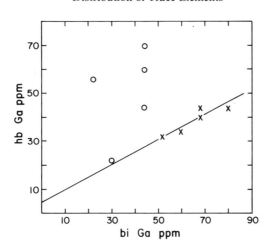

Fig. 22. Distribution of Ga between biotite and hornblende. Sample as in Fig. 20

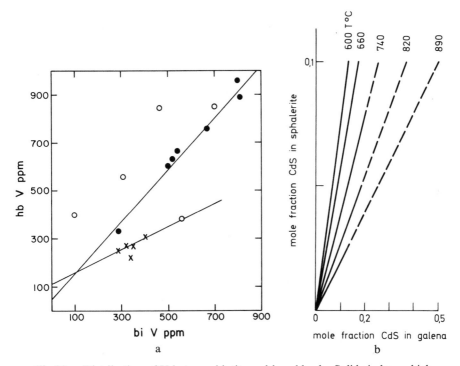

a   b

Fig. 23. a Distribution of V between biotite and hornblende. Solid circles — high amphibolite facies (KRETZ, 1959). Other symbols as in Fig. 20. b Experimentally determined distribution of CdS between sphalerite and galena at various temperatures. Data from BETHKE and BARTON (1971)

components in the minerals, it appears that Ni is enriched in hornblende relative to biotite with increasing temperature while V, Ga and Ba are enriched in biotite relative to hornblende.

BETHKE and BARTON (1971) have studied the distribution of elements such as cadmium, manganese and selenium between coexisting sulphide minerals in experimental systems. Their experiments show that the distribution coefficients are independent of composition and vary sufficiently with temperature to permit estimates of temperature of crystallization in suitable cases. Their data on the distribution of CdS between coexisting galena and sphalerite are shown in Fig. 23 b. The temperature dependence of the distribution coefficient for CdS between galena and sphalerite is given by

$$\log K_{CdS} = 2.080\,T^{-1} - 1.47\,.$$

Similar relationships are present in the distribution of manganese and selenium between the two minerals and in the distribution of cadmium and manganese between wurtzite and galena.

The results of the studies discussed above should encourage the study of temperature and compositional dependence of the distribution coefficients in experimental and natural mineral assemblages.

V. Measurement of Component Activity Using Composition of Coexisting Minerals

Experimental data on the distribution of a component between two coexisting crystalline solutions at a fixed P and T for systems such as olivine and pyroxene have been collected by NAFZIGER and MUAN (1967), LARIMER (1968), and MEDARIS (1969). Distribution data are also available for natural assemblages, but the P and T of their formation are indefinite. The distribution data from natural assemblages in many cases may be found to represent ion-exchange equilibrium closely. If precise P and T are not important, such data may be used to obtain useful information on the thermodynamic nature of mixing in the minerals. For this purpose, the thermodynamic equations according to various solution models for binary solutions presented in this chapter may be used.

The composition of coexisting phases that do not obey the same equation of state may be used to find the activity-composition relations in each phase in suitable cases. Ion-exchange between α and β with chemical formulae $(A, B) M$ and $(A, B) N$, respectively may be expressed as

$$A\text{-}\alpha + B\text{-}\beta \rightleftharpoons B\text{-}\alpha + A\text{-}\beta. \tag{V.1}$$

The equilibrium constant $K_{V.1}$ at a certain P and T is given by

$$K_{V.1} = \left(\frac{x_{B\text{-}\alpha}\, x_{A\text{-}\beta}}{x_{A\text{-}\beta}\, x_{B\text{-}\alpha}} \right) \left(\frac{f_{B\text{-}\alpha}\, f_{A\text{-}\beta}}{f_{A\text{-}\beta}\, f_{B\text{-}\alpha}} \right). \tag{V.2}$$

The term in the first bracket is the distribution coefficient K_D. Depending on the nature of the data available, the following cases may be considered.

1. Compositional Data Available on a Complete Distribution Isotherm

The simple mixture model, the two-constant asymmetric model, and the regular solution model with quasi-chemical approximation are the possible choices. According to the simple mixture model,

$$\ln K_{V.1} = \ln K_D - \frac{W_\alpha}{RT}(1 - 2x_{A\text{-}\alpha}) + \frac{W_\beta}{RT}(1 - 2x_{A\text{-}\beta}). \tag{IV.11}$$

A nonlinear least-squares fit using the data on $x_{A\text{-}\alpha}$ and $x_{A\text{-}\beta}$ finally yields $K_{V.1}$, W_α, and W_β.

According to the Redlich and Kister equations (KING, 1969, p. 326),

$$RT \ln f_A = x_B^2 [A_0 + A_1(3x_A - x_B) + A_2(x_A - x_B)(5x_A - x_B) + \cdots] \qquad (V.3)$$

and

$$RT \ln f_B = x_A^2 [A_0 - A_1(3x_B - x_A) + A_2(x_B - x_A)(5x_B - x_A) + \cdots] \qquad (V.4)$$

Therefore,

$$RT \ln \frac{f_A}{f_B} = A_0(x_B - x_A) + A_1(6x_A x_B - 1) + A_2(x_B - x_A)(1 - 8x_A x_B). \qquad (V.5)$$

Substituting the values of $f_{B\text{-}\alpha}/f_{A\text{-}\alpha}$ and $f_{A\text{-}\beta}/f_{B\text{-}\beta}$ by using Eq. (V.5) in Eq. (V.2), neglecting the constants A_2's, and rearranging in logarithmic form we have

$$\ln K_{V.1} = \ln K_D + \frac{A_{0\text{-}\alpha}}{RT}(x_{A\text{-}\alpha} - x_{B\text{-}\alpha}) + \frac{A_{1\text{-}\alpha}}{RT}(6x_{B\text{-}\alpha} x_{A\text{-}\alpha} - 1)$$
$$+ \frac{A_{0\text{-}\beta}}{RT}(x_{B\text{-}\beta} - x_{A\text{-}\beta}) + \frac{A_{1\text{-}\beta}}{RT}(6x_{A\text{-}\beta} x_{B\text{-}\beta} - 1). \qquad (V.6)$$

Eq. (V.6) is of the form

$$Y = C + A_1 x_1 + A_2 x_2 + A_3 x_3 + \cdots$$

where the Y and the x's are known quantities. It may be solved by a numeric least-squares method yielding $C = \ln K_{V.1}$ and other constants. There must be a minimum of five distribution points.

According to the quasi-chemical approximation,

$$f_A = \left[1 + \frac{\varphi_B(\beta - 1)}{\varphi_A(\beta + 1)} \right]^{z q_A/2}$$

and f_B is defined as in Eq. (II.10). Substituting these values of f into Eq. (V.2),

$$K_D \cdot \frac{\left[1 + \dfrac{\varphi_{A\text{-}\alpha}(\beta_\alpha - 1)}{\varphi_{B\text{-}\alpha}(\beta_\alpha + 1)} \right]^{z_\alpha q_{B\text{-}\alpha}/2} \left[1 + \dfrac{\varphi_{B\text{-}\beta}(\beta_\beta - 1)}{\varphi_{A\text{-}\beta}(\beta_\beta + 1)} \right]^{z_\beta q_{A\text{-}\beta}/2}}{\left[1 + \dfrac{\varphi_{B\text{-}\alpha}(\beta_\alpha - 1)}{\varphi_{A\text{-}\alpha}(\beta_\alpha + 1)} \right]^{z_\alpha q_{A\text{-}\alpha}/2} \left[1 + \dfrac{\varphi_{A\text{-}\beta}(\beta_\beta - 1)}{\varphi_{B\text{-}\beta}(\beta_\beta + 1)} \right]^{z_\beta q_{B\text{-}\beta}/2}} = K_{V.1} \qquad (V.7)$$

where β_α and β_β are for phases α and β, respectively, and are given by Eq. (II.11), with x_A replaced by φ_A, etc. q_α and q_β are contact factors for phases α and β, respectively. A numerical method would be required to solve Eq.(V.7). More details on calculations using quasi-chemical approximation are presented later in Chapters VI and VIII.

NAFZIGER and MUAN (1967) determined the activity-composition relation in pyroxene by using the composition of pyroxene and the oxygen pressure in the following reaction:

$$(FeSiO_3)_{\text{in pyroxene}} = \underset{\text{metal}}{Fe} + \underset{\text{silica}}{SiO_2} + \underset{\text{gas}}{1/2 O_2} \qquad (V.8)$$

For (V.8) we have

$$RT \ln a_{FeSiO_3} = \Delta G^0_{V.8} + 1/2\,RT \ln P_{O_2} \qquad (V.9)$$

where $\Delta G^0_{V.8}$ is the standard free energy change for the reaction, and P_{O_2} is the oxygen pressure of a gas phase in equilibrium with pyroxene crystalline solution, silica and metallic iron.

For expressing the activity coefficient in powers of x, the Eqs.(V.3, V.4) may be used. Substituting (V.3) in Eq.(V.9) and rearranging we have (with two constants):

$$RT \ln \frac{(P_{O_2})^{1/2}}{x_{FeSiO_3}} = -\Delta G^0_{V.8} + (1 - x_{FeSiO_3})^2 [A_0 + A_1(4 x_{FeSiO_3} - 1)] \qquad (V.10)$$

which may be solved by least-squares analysis as before. If the standard free energy change of the reaction can be determined independently from thermochemical data, the activity of a component in the crystalline solution can be calculated (Eq.V.9) without involving a solution model.

2. Composition Data on a Complete Distribution Isotherm and the Activity-Composition Relation in One of the Two Coexisting Phases

Depending on the accuracy of the data, Eqs.(IV.11), (V.6), and (V.7) may be used. If necessary, all three constants in Eq.(V.5) may be used. Eq.(V.5) may be written as

$$RT \ln \frac{f_A}{f_B} = -(A_0 + A_1 + A_2) + x_B(2A_0 + 6A_1 + 10A_2)$$
$$- x_B^2(6A_1 + 24A_2) + x_B^3(16A_2). \qquad (V.11)$$

Transforming Eq. (V.2) into logarithmic form and substituting values from Eq. (V.11) for $f_{A\text{-}\beta}/f_{B\text{-}\beta}$,

$$\ln K_{\mathrm{D}} + \ln \frac{f_{\mathrm{B}\text{-}\alpha}}{f_{\mathrm{A}\text{-}\alpha}} = \left(\ln K_{\mathrm{V.1}} + \frac{A_0 + A_1 + A_2}{RT} \right) - x_{\mathrm{B}} \left(\frac{2A_0 + 6A_1 + 10A_2}{RT} \right)$$
$$+ x_{\mathrm{B}}^2 \left(\frac{6A_1 + 24A_2}{RT} \right) - x_{\mathrm{B}}^3 \left(\frac{16A_2}{RT} \right). \tag{V.12}$$

Eq. (V.12) is of the form

$$Y = a_0 + a_1 x + a_2 x^2 + a_3 x^3 + \cdots$$

If the activity-composition relation in α is known, Eq. (V.12) may be solved by least-squares analysis.

If a sufficient number of distribution points is available, Eq. (V.6) may be used with the third constant A_2 and the results compared with those obtained by using Eq. (V.12).

In suitable cases, we may calculate the activity-composition relation in one of two coexisting phases, without involving a solution model. Consider the following reaction as presented by NAFZIGER and MUAN (1967)

$$\mathrm{FeSiO_3} + \mathrm{MgSi_{0.5}O_2} \rightleftharpoons \mathrm{MgSiO_3} + \mathrm{FeSi_{0.5}O_2}. \tag{V.13}$$
$$\text{px} \qquad \text{ol} \qquad\quad \text{px} \qquad\quad \text{ol}$$

The equilibrium constant for the reaction (V.13) is

$$K_{\mathrm{V.13}} = \frac{a_{\mathrm{En\text{-}px}}\, a_{\mathrm{Fa\text{-}ol}}}{a_{\mathrm{Fs\text{-}px}}\, a_{\mathrm{Fo\text{-}ol}}}$$
$$= \left(\frac{a_{\mathrm{En\text{-}px}}}{a_{\mathrm{Fs\text{-}px}}} \right) \left(\frac{x_{\mathrm{Fa\text{-}ol}}}{x_{\mathrm{Fo\text{-}ol}}} \right) \left(\frac{f_{\mathrm{Fa\text{-}ol}}}{f_{\mathrm{Fo\text{-}ol}}} \right) \tag{V.14}$$
$$= K'_{\mathrm{V.13}} \left(\frac{f_{\mathrm{Fa\text{-}ol}}}{f_{\mathrm{Fo\text{-}ol}}} \right)$$

where $K'_{\mathrm{V.13}}$ is $(a_{\mathrm{En\text{-}px}} x_{\mathrm{Fa\text{-}ol}})/(a_{\mathrm{Fs\text{-}px}} x_{\mathrm{Fo\text{-}ol}})$. Activity coefficients may be calculated from the expressions

and

$$\ln f_{\mathrm{Fa}} = - x_{\mathrm{Fo}} \ln K'_{\mathrm{V.13}} + \int_0^{x_{\mathrm{Fo}}} \ln K' \, d x_{\mathrm{Fo}} \tag{V.15}$$

$$\ln f_{\mathrm{Fo}} = - x_{\mathrm{Fa}} \ln K'_{\mathrm{V.13}} + \int_1^{x_{\mathrm{Fo}}} \ln K' \, d x_{\mathrm{Fo}}. \tag{V.16}$$

As $\ln K'_{\mathrm{V.13}}$ is known, the activity coefficients may be determined from Eqs. (V.15) and (V.16) by solving the integration graphically.

3. Examples of Calculations

Although the activity-composition relations in minerals may be calculated from the composition of the coexisting phases by using the equations mentioned above, the precision and reliability of such determinations depends on the accuracy of the measured composition data and on the attainment of equilibrium in the reaction in experimental or in the natural systems. Generally any solution model with two or more constants may be fitted to the compositional data but whether we can attach any physical significance to the energetic properties of the solution so determined, is something that has to be learnt by more experience and research on mineralogical systems.

SCHULIEN, FRIEDRICHSEN and HELLNER (1970) studied the distribution of Fe^{2+} and Mg^{2+} between a cloride solution (sol) and olivine at 723—923 °K. The nature of the chloride solution in the experiments is not clearly known. Assuming that this solution is a homogeneous liquid or gaseous solution, the data on the composition of the coexisting phases may be used to demonstrate the method of calculating activity-composition relation. It is the purpose of this example to show that the calculations of activities from Eqs. (IV.11, V.6, V.7, V.12 and V.14) do not always yield sensible values of the energy parameters A_0 and A_1 because these parameters are quite sensitive to the form of the isotherms. The following ion-exchange equilibrium may be assumed to have been established in the experiment (SCHULIEN, FRIEDRICHSEN and HELLNER, 1970):

$$\underset{\text{sol}}{FeCl_2} + \underset{\text{ol}}{MgSi_{0.5}O_2} \rightleftharpoons \underset{\text{sol}}{MgCl_2} + \underset{\text{ol}}{FeSi_{0.5}O_2} \qquad (V.17)$$

for which the equilibrium constant is

$$K_{V.17} = \left(\frac{x_{\text{Mg-sol}}\, x_{\text{Fe-ol}}}{x_{\text{Fe-sol}}\, x_{\text{Mg-ol}}} \right) \left(\frac{f_{\text{Mg-sol}}\, f_{\text{Fe-ol}}}{f_{\text{Fe-sol}}\, f_{\text{Mg-ol}}} \right). \qquad (V.18)$$

One of the three Eqs. IV.11, V.6 and V.7 may now be used. In the present case, calculations indicate that use of the Eqs. (V.6) or (V.7) does not yield thermodynamic parameters which vary consistently as a function of temperature. The W/RT values (simple mixture model) listed in Table 2 are also not very satisfactory since at 723 °K, W/RT exceeds 2.0 which would mean unmixing in olivines. Fig. 24 shows the calculated isotherms at 773 °K to 923 °K using the W/RT values in Table 2. The isotherm at 723 °K is drawn by using the linearly extrapolated values ($W_{ol}/RT = 2.0$, $W_{sol}/RT = 0.91$ and $K_{V.17} = 0.76$). W/RT for olivine at 923 °K is negligible showing that olivine may be closely ideal at this temperature. The nature of the olivine crystalline solution is discussed in another chapter.

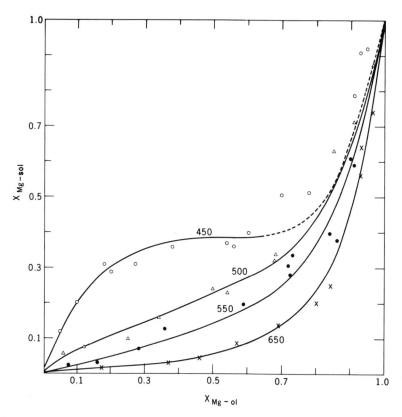

Fig. 24. Distribution of Mg^{2+}-Fe^{2+} between olivine and chloride solution. The data points are from SCHULIEN *et al.* (1970). Crosses 650° C, solid circles 550° C, triangles 500° C, and open circles 450° C. The curves are least squares curves using the "simple mixture" model for both the solutions

Table 2. Interchange energy W and the standard free energy change ΔG^0 in olivine-chloride solution system

T(K)	W_{sol}/RT	W_{ol}/RT	$-RT \ln K = \Delta G^0_{V.17}$ (cal/mole)
723	1.86	2.32	1433
773	0.87	1.28	723
823	0.59	0.73	421
923	0.52	−0.014	181

Table 3. Composition of coexisting cummingtonite $(Fe,Mg)_7Si_8O_{22}(OH)_2$ and actinolite $(Fe,Mg)_5Ca_2Si_8O_{22}(OH)_2$ in iron formations (see also KISCH and WARNAARS, 1969)

$x_{Fe\text{-}cum}$	$x_{Mg\text{-}cum}$	$x_{Fe\text{-}act}$	$x_{Mg\text{-}act}$	$K_{D(V.21)}$	Ref.
0.783	0.217	0.714	0.286	1.45	MUELLER (1960)
0.668	0.332	0.560	0.440	1.58	MUELLER (1960)
0.652	0.348	0.547	0.453	1.55	KLEIN (1968)
0.580	0.420	0.427	0.573	1.84	MUELLER (1960)
0.457	0.543	0.307	0.693	1.90	MUELLER (1960)
0.230	0.770	0.129	0.871	2.03	MUELLER (1960)

Calculation of activities are more reliable from the compositional data if there is a systematic variation in the distribution coefficient K_D with the composition of the phases. Consider, for example, the data presented in Table 3 which show the composition of coexisting cummingtonite and actinolite in certain iron-formations. The ion-exchange reaction may be expressed as

$$\tfrac{1}{7}Mg_7Si_8O_{22}(OH)_2 + \tfrac{1}{5}\underset{\text{Fe-act}}{Fe_5Ca_2Si_8O_{22}(OH)_2} \rightleftarrows \tfrac{1}{7}\underset{\text{Fe-cum}}{Fe_7Si_8O_{22}(OH)_2}$$

$$+ \tfrac{1}{5}\underset{\text{Mg-act}}{Mg_5Ca_2Si_8O_{22}(OH)_2} \qquad (V.19)$$

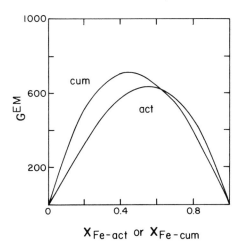

Fig. 25. Excess free energy of mixing (cal/mole) in cummingtonite and actinolite to compare the asymmetric character of the solutions

for which we may write

$$\ln K_{D(V.19)} = \ln K_{V.19} - \frac{A_{0\text{-act}}}{RT}(x_{\text{Fe-act}} - x_{\text{Mg-act}}) - \frac{A_{1\text{-act}}}{RT}(6x_{\text{Fe-act}}$$

$$(V.20)$$

$$\cdot\, x_{\text{Mg-act}} - 1) - \frac{A_{0\text{-cum}}}{RT}(x_{\text{Mg-cum}} - x_{\text{Fe-cum}}) - \frac{A_{1\text{-cum}}}{RT}(6x_{\text{Fe-cum}}\, x_{\text{Mg-cum}} - 1)$$

where $K_{D(V.19)}$ is $(x_{\text{Mg-act}}\,x_{\text{Fe-cum}})/(x_{\text{Fe-act}}\,x_{\text{Mg-cum}})$. A multiple regression analysis of Eq.(V.20) yields $A_{0\text{-act}}/RT$ and $A_{1\text{-act}}/RT$ as 1.63 and 0.44 respectively and $A_{0\text{-cum}}/RT$ and $A_{1\text{-cum}}/RT$ as 1.86 and -0.38 respectively. $\ln K$ is 0.109. Note that in Eq.(V.20), component 1 for actinolite is Mg-actinolite and for cummingtonite it is Fe-cummingtonite. Therefore, although A_1 is negative in one case and positive in the other yet both the crystalline solutions are asymmetric in the same way. Fig. 25 shows the excess free energy of mixing G^{EM} plotted against $x_{\text{Fe-act}}$ or $x_{\text{Fe-cum}}$. Figs. 26 and 27 show the activity-composition relations in actinolite and cummingtonite crystalline solutions assuming that the temperature of ion-exchange equilibrium is close to 500° C (see MUELLER, 1960). These activity calculations cannot be quantitative because they depend not only on the accuracy of the compositional data, but also on the correctness of the assumptions that equilibrium was closely approached

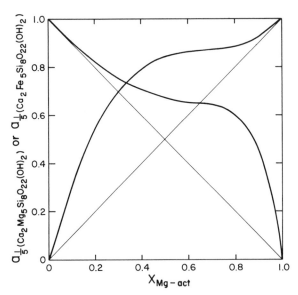

Fig. 26. Activity of Mg and Fe end members in actinolite series plotted against mole fraction of Mg in actinolite at ∼ 500° C. The results are only qualitative

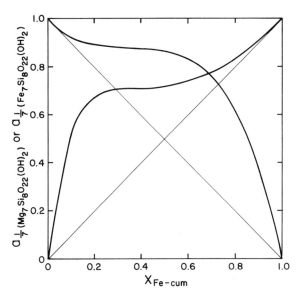

Fig. 27. Activity of Mg and Fe end members in cummingtonite plotted against mole fraction of Fe in cummingtonite at $\sim 500°$ C. The results are only qualitative

in all the samples at the same P and T and that the concentration of components other than Fe and Mg have a negligible effect on the energetic properties of the crystalline solutions. The deviations from ideality shown in Figs. 26 and 27 appear to be somewhat too large. Such results should be tested by considering the composition of these minerals with some other coexisting minerals with known or similarly inferred activity-composition relations, such as orthopyroxene. Such efforts may finally yield activity-composition data for the quasi-binary Fe–Mg silicates which are internally consistent.

VI. Measurement of Component Activities by Analysis of Two-Phase Data

An experimental measurement of activities of components in a crystalline solution, particularly the silicates, is beset with difficulties and the measured values are subject to large errors. Therefore, obtaining such activity-composition relations from phase diagrams would be very convenient.

However, there is no direct method of doing this without some kind of a solution model. The use of a solution model brings in uncertainties in the activity values, which depend in extent and form on the choice of the model. The attempts to obtain the thermodynamic functions of mixing through the use of various solution models is still useful. For some crystalline solutions it may be possible to experimentally determine such properties. A comparison between the experimental values and the values based on a particular model would provide a greater understanding of the interrelationship of the crystal structural parameters on which the model is based and the thermodynamics of the crystal phase. In other cases where experimental determinations cannot be made, the empirically derived functions of mixing may be tested for their physical significance by their success in application to problems of petrogenesis.

1. Symmetrical Mixtures

a) Simple Mixture

The thermodynamics of a simple mixture or regular symmetric solution has been discussed before. Assume that the solution (A, B) unmixes into two coexisting solutions, α, rich in A, and β, rich in B. At equilibrium,

$$\mu_{A\text{-}\alpha} = \mu_{A\text{-}\beta}$$

and
$$\mu_{B\text{-}\alpha} = \mu_{B\text{-}\beta} \tag{VI.1}$$

so that according to the simple mixture model,

$$\mu^0_{A\text{-}\alpha} + RT \ln x_{A\text{-}\alpha} + W(1 - x_{A\text{-}\alpha})^2 = \mu^0_{A\text{-}\beta} + RT \ln x_{A\text{-}\beta} + W(1 - x_{A\text{-}\beta})^2 \tag{VI.2}$$

where μ^0's stand for the same pure end member structure A and can be cancelled. Substituting $x_{A-\beta} = 1 - x_{A-\alpha}$,

$$RT \ln x_{A-\alpha} + W(1 - x_{A-\alpha})^2 = RT \ln(1 - x_{A-\alpha}) + W(x_{A-\alpha})^2 \qquad \text{(VI.3)}$$

or

$$\frac{W}{RT} = \frac{\ln[(1 - x_{A-\alpha})/x_{A-\alpha}]}{1 - 2x_{A-\alpha}}. \qquad \text{(VI.4)}$$

This expression is similar to the one obtained by THOMPSON (1967). The equation for the curve of coexistence of two phases may also be written in terms of critical temperature T_c of unmixing and the mole fractions by substituting

$$W = 2RT_c$$

into Eq. (VI.4):

$$T = 2T_c \frac{1 - 2x_{A-\alpha}}{\ln[(1 - x_{A-\alpha})/x_{A-\alpha}]}. \qquad \text{(VI.5)}$$

If there are data on the composition of coexisting phases at different temperatures and the form of the solvus is symmetric, the value of W and the activity-composition relations can be calculated.

b) Symmetrical Mixture of Higher Order

Symmetrical crystalline solutions may not be simple mixtures and may require an expression with two or more constants to represent G^{EM}:

$$G^{EM} = x_{A-\alpha}(1 - x_{A-\alpha})[A_0 + A_2(1 - 2x_{A-\alpha})^2]. \qquad \text{(VI.6)}$$

For such a solution, Eq. (VI.3) is

$$RT \ln x_{A-\alpha} + [A_0 + A_2(1 - 2x_{A-\alpha})^2](1 - x_{A-\alpha})^2 = RT \ln x_{A-\beta}$$
$$+ [A_0 + A_2(1 - 2x_{A-\beta})^2](1 - x_{A-\beta})^2. \qquad \text{(VI.7)}$$

Using the relation $\mu_{B-\alpha} = \mu_{B-\beta}$, an equation similar to Eq. (VI.7) can be written, and the two equations can be solved simultaneously to obtain A_0, A_2, and the activity-composition relation.

2. Asymmetrical Solutions

a) Subregular Model

As mentioned before, G^{EM} may be expressed as a polynomial in the mole fraction x_A or x_B for the compound $(A, B)M$ according to Guggenheim's equation:

$$G^{EM} = x_A x_B [A_0 + A_1(x_A x_B) + A_2(x_A - x_B)^2 + \cdots]. \qquad \text{(II.21)}$$

If $A_2 = 0$ is substituted into Eq. (II.21), a two-constant equation for an asymmetrical solution is the result. Proceeding as in the previous sections,

$$RT \ln x_{A-\alpha} + RT \ln f_{A-\alpha} = RT \ln x_{A-\beta} + RT \ln f_{A-\beta} \qquad (VI.8)$$

and

$$RT \ln x_{B-\alpha} + RT \ln f_{B-\alpha} = RT \ln x_{B-\beta} + RT \ln f_{B-\beta} \qquad (VI.9)$$

Substituting for $RT \ln f$ from Eqs. (II.22) and (II.23),

$$RT \ln x_{A-\alpha} + (x_{B-\alpha}^2)\{A_0 + A_1(3x_{A-\alpha} - x_{B-\alpha})\}$$
$$= RT \ln x_{A-\beta} + (x_{B-\beta}^2)\{A_0 + A_1(3x_{A-\beta} - x_{B-\beta})\} \qquad (VI.10)$$

and

$$RT \ln x_{B-\alpha} + (x_{A-\alpha}^2)\{A_0 - A_1(3x_{B-\alpha} - x_{A-\alpha})\}$$
$$= RT \ln x_{B-\beta} + (x_{A-\beta})^2\{A_0 - A_1(3x_{B-\beta} - x_{A-\beta})\} \qquad (VI.11)$$

'Eqs. (VI.10) and (VI.11) can now be solved simultaneously to yield the values of the two constants A_0 and A_1. This method of calculation is equivalent to that used by THOMPSON (1967) and THOMPSON and WALD-BAUM (1969a, b) as may be noted by substituting for A_0 and A_1 from Eqs. (II.34a, b) (see Eqs. (II.26) to (II.33) for various expressions). By calculating A_0 and A_1 from the binodal solvus data obtained at different P and T, it is possible to determine the thermodynamic properties of the crystalline solution such as free energy of mixing (II.21) heat (II.25) and entropy of mixing (II.24) and the activity coefficients (II.22 and II.23). The binodal bounding the two-phase region may be determined graphically by the double tangent method on a plot of free energy of mixing against mole fraction or by a numerical iteration method which solves Eqs. (VI.10) and (VI.11) with the mole fractions as unknown quantities (see program Subroutine Bigap in Appendix).

b) Quasi-Chemical Approximation

This solution model has been discussed in Chapter II. GREEN (1970) used this model to study the binodal data on the halite-sylvite (NaCl–KCl) system. Let us consider the crystalline solution (A, B) which unmixes into two coexisting solutions α, rich in A, and β, rich in B. The conditions of equilibrium are given as before by Eqs. (VI.1, VI.8, and VI.9).

Substituting values of f_A and f_B from Eqs. (II.16) and (II.17), respectively, into Eqs. (VI.8) and (VI.9),

$$\ln x_{A-\alpha} + \frac{z_\alpha q_{A-\alpha}}{2} \ln \left[1 + \frac{\varphi_{B-\alpha}(\beta - 1)}{\varphi_{A-\alpha}(\beta + 1)}\right] = \ln x_{A-\beta} + \frac{z_\beta q_{B-\beta}}{2} \left[1 + \frac{\varphi_{B-\beta}(\beta' - 1)}{\varphi_{A-\beta}(\beta' + 1)}\right]$$
$$(VI.12)$$

$$\ln x_{B\text{-}\alpha} + \frac{z_\alpha q_{A\text{-}\alpha}}{2} \ln \left[1 + \frac{\varphi_{A\text{-}\alpha}(\beta - 1)}{\varphi_{B\text{-}\alpha}(\beta + 1)} \right] = \ln x_{B\text{-}\beta} + \frac{z_\beta q_{B\text{-}\beta}}{2} \left[1 + \frac{\varphi_{A\text{-}\beta}(\beta' - 1)}{\varphi_{B\text{-}\beta}(\beta' + 1)} \right]$$

(VI.13)

where β and β' correspond to phases α and β, respectively, and the ϕ's are defined by Eqs. (II.18). q_A and q_B are the contact factors discussed before. They are not independent and should approach 1 simultaneously. GREEN (1970) assumed $q_A q_B = 1$. The two independent relations (VI.12) and (VI.13) contain two unknowns q_A/q_B and W and can be solved by an iteration process. The ratio q_A/q_B is a function of the geometry of the substituting chemical species and therefore may be regarded as almost independent of T. Substitution of q_A/q_B back into Eqs. (VI.12) and (VI.13) gives two independent values of W at each temperature. Any difference noted in the two values of W would be caused by the inadequacy of the solution model to fit to the experimental data.

The solvus bounding the two-phase region may be determined graphically by the double tangent method on a plot of free energy of mixing against mole fraction or by a numerical iteration method. The excess functions of mixing can be calculated by Eqs. (II.19) and (II.20).

3. Example of Calculation of Functions of Mixing: The CaWO$_4$–SrWO$_4$ System

CHANG (1967) presented two-phase data for various binary tungstate ($R^{II}WO_4$) crystalline solutions. To demonstrate the method of calculation, the compositions of the coexisting phases are determined from Fig. 28 (Fig. 3, CHANG, 1967) and tabulated in Table 4. Calculated values of A_0/RT and A_1/RT according to the subregular model are listed in Table 5. The relationship between A_0/RT and A_1/RT and T is linear and

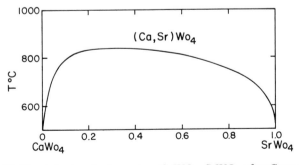

Fig. 28. Binodal solvus in the system CaWO$_4$–SrWO$_4$ after CHANG (1967)

Table 4. Chemical composition of unmixed phases in the system $CaWO_4$–$SrWO_4$

$T(°K)$	$x_{SrWO_4-\alpha}$	$x_{CaWO_4-\alpha}$	$x_{SrWO_4-\beta}$	$x_{CaWO_4-\beta}$
823	0.005	0.995	0.995	0.005
873	0.010	0.990	0.980	0.020
923	0.025	0.975	0.955	0.045
973	0.035	0.965	0.905	0.095
1023	0.120	0.880	0.630	0.370

α and β are coexisting phases rich in $CaWO_4$ and $SrWO_4$, respectively. The compositions are from CHANG (1967, Fig. 3).

Table 5. The calculated A_0/RT and A_1/RT in $(Ca, Sr)WO_4$

$T°(K)$	A_0/RT	A_1/RT
600	4.364	−0.327
650	3.524	−0.348
700	3.119	−0.455
750	2.537	−0.514
800	2.021	−0.607
900[a]	0.839	−0.739
1000[a]	−0.296	−0.884

[a] From Eqs.(VI.14) and (VI.15). Note: An error of $\pm 5\%$ in the mole fractions (Table 4) results in a ± 200 cal/mole error in determining A_0 and A_1.

is given by

$$A_0/RT = 14.1526 - 0.01135\,T \qquad (VI.14)$$

and

$$A_1/RT = 0.9616 - 0.00145\,T \qquad (VI.15)$$

G^{EM} can then be calculated from the relation

$$G^{EM} = x_{SrWO_4} x_{CaWO_4} [A_0 + A_1 (x_{SrWO_4} - x_{CaWO_4})],$$

and the activity coefficients from the relations

$$RT \ln f_{SrWO_4} = x^2_{CaWO_4} [A_0 + A_1 (3\,x_{SrWO_4} - x_{CaWO_4})]$$

and

$$RT \ln f_{CaWO_4} = x^2_{SrWO_4} [A_0 - A_1 (3\,x_{CaWO_4} - x_{SrWO_4})].$$

Fig. 29 shows the activity-composition relation at 1073 and 1273 °K. G^{EM} can also be plotted against composition, and the composition of the coexisting phases can be found by the tangent method. In the present case, the differences between compositions calculated by the model and the observed compositions in Table 4 are found to be small.

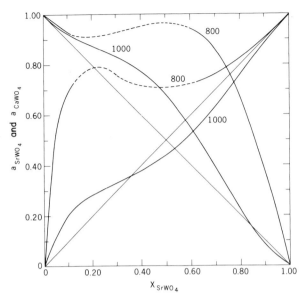

Fig. 29. Activity-composition relation in CaWO$_4$–SrWO$_4$ crystalline solution at 800 and 1000° C

As mentioned before, the activity-composition data and other thermodynamic functions as calculated from phase diagrams are sensitive to the nature of the assumptions and the model used. For the system CaWO$_4$–SrWO$_4$, calculation shows that the use of the quasi-chemical approximation predicts the solvus with the same accuracy as the subregular model does. The use of quasi-chemical approximation requires the values of q_A, q_B, and the co-ordination number z, the number of the nearest Ca^{2+} or Sr^{2+} ions surrounding each other. In CaWO$_4$ there are four Ca^{2+} ions surrounding each Ca^{2+} at a distance of approximately 3.9 Å. There are four more Ca^{2+} ions at a distance of 5 Å. z may be assumed equal to 4, and the two equations of the quasi-chemical approximation (Eqs. VI.12 and VI.13) can be solved simultaneously to find q_A/q_B.

In this case, it may be assumed that either $q_A + q_B = 2$ or $q_A q_B = 1$. The differences in the calculations of W using either of the two assumptions are small (see GREEN, 1970). A computer program may be used to solve each of the two equations independently by using various values for q_A and q_B and to compare the W values so obtained until the best set of W values is found. The W values with $z = 4$ are listed in Table 6. q_A and q_B are 1.20 and 0.80, corresponding to SrWO$_4$ and

Table 6. $2W/zRT$ for the system
CaWO$_4$–SrWO$_4$

$T°$(K)	Eqs.(VI.12)	Eq.(VI.13)
873	5.883	5.361
923	5.173	4.821
973	4.919	4.460
1023	4.495	4.301
1073	4.183	4.386
$z=4$; $q_1=1.20$; $q_2=0.80$		

CaWO$_4$, respectively. The atomic radii for Sr^{2+} and Ca^{2+} are 1.12 and 0.99 respectively (AHRENS, 1952). The ratio of q_A to q_B is 1.50 and the ratio of the radii of Sr^{2+} to Ca^{2+} is 1.13; these values are not similar.

The following equation describes the relation between the calculated W (the average of the two values listed in Table 6) and T:

$$2 W/zRT = 12.47258 - 0.00796 \, T \qquad (VI.16)$$

Fig. 30 shows a comparison of G^{EM} at 1073 °K calculated according to both the subregular (SR) model and the quasi-chemical (QC) model. The value of G^{EM} according to the latter is nearly twice that calculated according to the former. Differences between the other calculated functions of mixing, H^{EM} and S^{EM}, are even more marked. Unfortunately there are no data on experimentally determined H^{EM} and S^{EM} for the CaWO$_4$–SrWO$_4$ system, and, therefore, there is no way to know which model predicts the thermodynamic functions better in this particular case.

For the system NaCl–KCl, GREEN (1970) compared the thermodynamic quantities calculated by the subregular model and the quasi-chemical model with those determined by experiments. The thermodynamic quantities predicted by the quasi-chemical model are closer to those measured experimentally.

A comparison of the predictions of the functions of mixing in several binary alloys by the regular solution model and by the quasi-chemical model (LUPIS and ELLIOTT, 1967) shows that generally the predictions by the latter for the excess free energy are closer in agreement with experimental determinations than those by the former. The prediction for the excess entropy from the quasi-chemical model is not satisfactory. This may be in part caused by the neglect of the nonconfigurational excess entropy in many of the binary alloys. For the halite-sylvite system, GREEN (1970) finds that the nonconfigurational contributions are unimportant and suggests that the positive excess entropy of mixing found

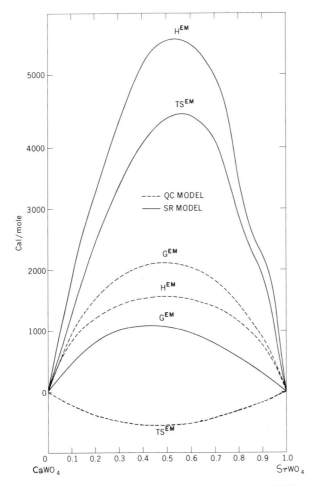

Fig. 30. Thermodynamic excess functions of mixing in (Ca,Sr)WO$_4$ according to two constant asymmetric model and according to quasi-chemical approximation

in the NaCl–KCl system may result from the introduction of vacancies or other defects into a crystalline solution.

This approach of calculation thermodynamic functions of mixing by the analysis of phase diagrams is relatively new in the field of mineralogy and deserves more attention from mineralogists and petrologists. The fact, that there is no unique analysis of a solvus and two or more solution models may be applicable to the same solvus data, need not be a barrier to acquiring and interpreting more phase data with the help of various solution models. Experimental verification of many of these results may not be possible in the near future. However, it may be possible to test such thermodynamic data by the success of their application to petrologic problems.

4. Coexisting Phases with Different Crystal Structures

In minerals such as pyroxenes and amphiboles, one often finds a pair of minerals whose composition lies on a binary compositional join but whose structures differ somewhat from each other. A case in point is the enstatite-diopside pair. The composition in this series lies on the binary $MgSiO_3$–$CaSiO_3$ join. The solvus on this join is not comparable to the solvus for the NaCl–KCl system or for the $CaWO_4$–$SrWO_4$ system. The important difference is that enstatite and diopside do not obey the same equations of state as do NaCl and KCl or $CaWO_4$ and $SrWO_4$. Phase equilibrium data on the enstatite-diopside solvus such as that of BOYD and SCHAIRER (1964), therefore, cannot be used for evaluating the activity-composition relations by the methods of previous sections.

Consider two coexisting phases α and β with different structures but the same composition (A, B) M. Then,

$$\mu_{AM-\alpha} = \mu_{AM-\beta} \tag{VI.17}$$

and

$$\mu_{BM-\alpha} = \mu_{BM-\beta} \tag{VI.18}$$

where

$$\mu_{AM-\alpha} = \mu^0_{AM-\alpha} + RT \ln x_{AM-\alpha} + W_\alpha (1 - x_{AM-\alpha})^2 \tag{VI.19}$$

and the other μ's are similarly defined.

Substituting Eq. (VI.19) into Eqs. (VI.17) and (VI.18) we have

$$\mu^0_{AM-\alpha} + RT \ln x_{AM-\alpha} + W_\alpha (1 - x_{AM-\alpha})^2 = \mu^0_{AM-\beta} + RT \ln x_{AM-\beta}$$
$$+ W_\beta (1 - x_{AM-\beta})^2 \tag{VI.20}$$

and

$$\mu^0_{BM-\alpha} + RT \ln x_{BM-\alpha} + W_\alpha (1 - x_{BM-\alpha})^2 = \mu^0_{BM-\beta} + RT \ln x_{BM-\beta}$$
$$+ W_\beta (1 - x_{BM-\beta})^2 . \tag{VI.21}$$

Because α and β do not have the same structure and, therefore, do not obey the same equation of state, generally $\mu^0_{AM-\alpha} \neq \mu^0_{AM-\beta}$ and $\mu^0_{BM-\beta} \neq \mu^0_{BM-\alpha}$. Therefore, the values of these chemical potentials at given P and T must be known to be able to solve Eqs. (VI.20) and (VI.21) simultaneously to find W_α and W_β. The free energy diagram in Fig. 31 shows this situation. In a binary system at a given P and T, two coexisting phases with compositions representing the minimum free energy may be present.

The enstatite-diopside system may be similar to the example shown in Fig. 31. The binary join is $(Mg, Mg)Si_2O_6$–$(Ca, Mg)Si_2O_6$. There are

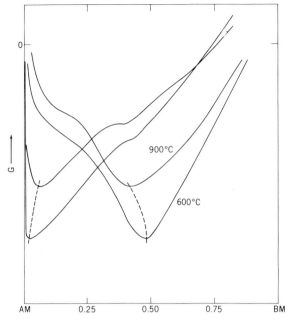

Fig. 31. Free energy of mixing in two crystalline solutions with the same chemical components AM and BM but with different crystal structures. The enstatite-diopside system is similar in principle to this hypothetical system

further complications because $(Ca, Mg)Si_2O_6$ with orthopyroxene structure and $(Mg, Mg)Si_2O_6$ with diopside structure are unknown.

In general structural information is not necessary for thermodynamic calculations. When available, however, such information may be used to interpret the observed thermodynamic behaviour of a crystalline solution. As long as there is a complete crystalline solution between two end-members, their structures need not be considered as different for the purpose of these calculations. Indeed, it is the purpose of thermodynamic calculations to find how the total free energy changes as a function of mixing of the two species, regardless of their structures.

Thus, to give another example, it is formally correct to calculate the activity-composition relation in the system Co_2SiO_4–Zn_2SiO_4 as long as there is no unmixing. But when unmixing occurs, the following problem arises. Let α and β be the two coexisting phases — α rich in Zn_2SiO_4 and β rich in Co_2SiO_4 at a certain temperature. At equilibrium,

$$\mu_{Co_2SiO_4\text{-}\alpha} = \mu_{Co_2SiO_4\text{-}\beta} \qquad (VI.22)$$

where $\mu_{Co_2SiO_4}$ is the chemical potential of the component Co_2SiO_4 in the solutions α and β. The structure of α [willemite] is different from that of

β (olivine) because Co occupies tetrahedral sites in α and octahedral sites in β. Therefore we have:

$$\mu_{Co_2SiO_4\text{-}\alpha} = \mu^0_{[Co_2SiO_4]} + RT \ln a_{Co_2SiO_4\text{-}\alpha} \qquad (VI.23)$$

and

$$\mu_{Co_2SiO_4\text{-}\beta} = \mu^0_{(Co_2SiO_4)} + RT \ln a_{Co_2SiO_4\text{-}\beta} \qquad (VI.24)$$

where $\mu^0_{[Co_2SiO_4]}$ is the chemical potential of Co_2SiO_4 with the willemite structure and $\mu^0_{(Co_2SiO_4)}$ is the chemical potential of Co_2SiO_4 with the olivine structure. Now,

$$\mu^0_{[Co_2SiO_4]} \neq \mu^0_{(Co_2SiO_4)} \qquad (VI.25)$$

and, therefore, the pure potentials do not cancel in Eq. (VI.22). From similar arguments $\mu^0_{[Zn_2SiO_4]} \neq \mu^0_{(Zn_2SiO_4)}$. In this situation it may be useful to introduce "fictive" end-members to explain activity-composition relations on either side of the solvus.

VII. Order-Disorder in Silicates

Long range order-disorder phenomena in silicates differ from those in alloys in several important respects. First, as opposed to alloys, silicates contain structural sites with definite polyhedral geometries and order-disorder is noticed whenever an ion occupies two or more sites which differ in their polyhedral shape and size. Second, usually only the ions which occupy the nonequivalent sites, take part in the order-disorder and the remaining silicate framework remains more or less inert. Third, when the ions are sufficiently alike in their charge and size, the site preference energies (corresponding to the difference in binding energy of the ion between the nonequivalent sites) are not strongly dependent on the degree of order as is usual in many binary alloys. Fourth, complete order or disorder in silicates is generally not possible because the variation in composition in binary solution would require that some of the ions of one species must inevitable occupy some of the sites belonging to the other species and also because there may be potential barrier for the ordering process to continue below a certain temperature. At the high temperature side there is usually a phase transformation or melting before complete disorder can be attained.

Order-disorder or the intracrystalline cation distribution in silicates is measurable by X-ray (see GHOSE, 1961) and other spectroscopic techniques. The energy of the intracrystalline ion exchange is part of the Gibbs free energy of the crystal and is, therefore, a very useful thermodynamic quantity (see MUELLER, 1969; THOMPSON, 1969).

Before considering order-disorder in crystalline solutions with two or more nonequivalent crystal structural sites, let us consider order-disorder in solutions with all sites structurally similar to each other. For such solutions the thermodynamic relations are simple and well understood.

1. Order-Disorder and the Crystalline Solution Models

As mentioned before the entropy of mixing in the regular solution model with the zeroth approximation is the same as that of the ideal solution. In other words it is assumed that there is complete disorder in the crystalline solution. Following GUGGENHEIM (1952) let a binary mixture contain $N_A = N(1-x)$ atoms of A and $N_B = Nx$ atoms of B on a lattice of N sites with a coordination number z. There are in all

$\frac{1}{2} z N$ pairs of closest neighbours namely A A, B B and A B. The assumption of complete randomness implies the conversion of all A A and B B pairs to A B pairs whose number is proportional to $\bar{x}$ given by

$$\bar{x} = \frac{N_A N_B}{N_A + N_B}.$$ (VII.1)

The interaction energy w (see Chapter II) is given by

$$2 w = 2 w_{AB} - w_{AA} - w_{BB}$$ (II.2)

The presence of interaction energy implies that the mixing of A and B cannot be really random. In the quasi-chemical approximation the average value $\bar{x}$ is given by

$$\bar{x}^2 = (N_A - \bar{x})(N_B - \bar{x}) \exp(-2w/z K T)$$ (VII.2)

Multiplying w by N (Avogadros number) and using the symbol W for Nw and the symbol η for $\exp(W/zRT)$, we may write

$$(N_A - \bar{x})(N_B - \bar{x}) - \eta^2 \bar{x} = 0.$$ (VII.3)

If $\bar{x}$ is expressed as

$$\bar{x} = \frac{N_A N_B}{N_A + N_B} \frac{2}{\beta + 1}$$ (VII.4)

it is possible to write for the quantity β from Eqs. (VII.3) and (VII.4)

$$\beta = \{1 + 4x(1 - x)(\eta^2 - 1)\}^{1/2}$$ (II.11)

where x is the mole fraction $N_A/N_A + N_B$. Multiplying both sides of Eq. (VII.4) by the total number of atoms, we have,

$$\bar{x}' = x(1 - x)(2/(\beta + 1))$$ (VII.5)

where $\bar{x}'$ is $\bar{x}(N_A + N_B)$. $\beta = 1$ corresponds to complete disorder since we now have $\bar{x}' = x(1 - x)$. Substituting $\beta = 1$ in (II.2) gives η or exp $(W/zRT) = 1$. In such a case the ideal solution model is obtained for which $W = 0$, β or $\eta = 1$.

Positive deviations from ideal solution occur if $W > 0$, which would mean $\eta < 1$, $\beta > 1$, and, therefore $\bar{x}' < x(1 - x)$. The number of AB pairs is less than that of the completely random configuration. This leads to a positive excess entropy of mixing besides the positive enthalpy. $W > 0$ therefore indicates a trend towards clustering and eventual un-mixing into two phases.

Negative deviations from ideal solution occur when $W < 0$ and there-fore $\eta > 1$, $\beta < 1$ and $\bar{x}' > x(1 - x)$. In such solutions there is a trend towards compound formation and the free energy of mixing is more nega-

Table 7. Data on relationship between mole fraction, ordering and quasi-chemical solution parameter

x_{SrWO_4}	0.10	0.30	0.50	0.70	0.90
x_{CaWO_4}	0.90	0.70	0.50	0.30	0.10
ϕ_{SrWO_4}	0.143	0.391	0.600	0.778	0.931
ϕ_{CaWO_4}	0.857	0.609	0.400	0.222	0.069

$2W/zRT = 5.622$, $z = 4$, $T = 873\,°K$

β	11.54	16.23	16.29	13.83	8.46
$\bar{x}'$	0.014	0.024	0.029	0.028	0.019
$\bar{x}'/x$	0.144	0.081	0.058	0.041	0.021

$2W/zRT = 3.135$, $z = 4$, $T = 1173\,°K$

β	3.43	4.68	4.70	4.03	2.58
$\bar{x}'$	0.040	0.074	0.088	0.084	0.050
$\bar{x}'/x$	0.406	0.246	0.175	0.119	0.056

$2W/zRT = -1.640$, $z = 4$, $T = 1773\,°K$

β	0.78	0.48	0.47	0.67	1.00
$\bar{x}'$	0.101	0.283	0.339	0.252	0.090
$\bar{x}'/x$	1.012	0.945	0.678	0.360	0.100

tive than that of an ideal solution. YAROSHEVSKIY (1970) has discussed the possibility of negative deviations from ideality in certain silicate solutions.

It is instructive to consider the interrelationship of order-disorder and the various quasi-chemical solution parameters by defining a quasi-chemical ordering parameter $Q(\bar{x}/x)$. The data on the various parameters are presented in Table 7, and Fig. 32 shows a plot of such an ordering parameter against the mole fraction x_{SrWO_4} in the $(Ca, Sr)WO_4$ crystalline solution. The quantities $2W/zRT$ at different temperatures are calculated from Eq. (VI.16). W is 0 at $1567\,°K$ and is negative at $1773\,°K$. The excess entropy of mixing is negative at lower temperatures and approaches zero at $1567\,°K$. There is random mixing at $1567\,°K$ and above this temperature. As the temperature is lowered the trend towards ordering increases and below the critical temperature the crystalline solution unmixes into two phases.

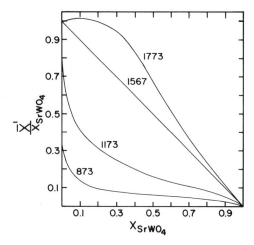

Fig. 32. Dependence of ordering on the composition and the quasi-chemical solution parameter W in binary $(Ca, Sr)WO_4$ solution. Numbers refer to $T°K$. The $2W/zRT$ values (Eq. VI.16) are 5.622, 3.135, 0 and -1.640 at 873, 1173, 1567 and 1773 °K respectively

Order-disorder in the crystalline solution with all sites equivalent as discussed above can be interpreted in terms of simple solution models. When there are more than one structural site, the treatment according to classical thermodynamics is somewhat complicated as discussed below.

2. Intracrystalline Ion Exchange and Site Activities

A crystalline solution $(A_x, B_{1-x})M$ may have two cations A and B distributed between two nonequivalent sites α and β. M is the inert silicate framework. Following DIENES (1955) and MUELLER (1960, 1962, 1969), the disordering process may be represented by the following exchange reaction:

$$A\text{-}\alpha + B\text{-}\beta \rightleftharpoons A\text{-}\beta + B\text{-}\alpha . \tag{VII.6}$$

In terms of kinetic theory, the time rate of change of A in site β is given by

$$-\frac{dx_{A\text{-}\beta}}{dt} = K_{\beta\text{-}\alpha}\,\varphi_{\beta\text{-}\alpha}\,x_{A\text{-}\beta}\,x_{B\text{-}\alpha} - K_{\alpha\text{-}\beta}\,\varphi_{\alpha\text{-}\beta}\,x_{B\text{-}\beta}\,x_{A\text{-}\alpha} \tag{VII.7}$$

where x refers to the mole fractions, $K_{\beta\text{-}\alpha}$ and $K_{\alpha\text{-}\beta}$ are rate constants and functions of P and T only, and $\varphi_{\beta\text{-}\alpha}$ and $\varphi_{\alpha\text{-}\beta}$ are analogous to activity coefficient products in a macroscopic chemical system and are functions of P, T, and composition. At equilibrium,

$$\frac{dx_{A\text{-}\beta}}{dt} = 0$$

and   (VII.8)

$$K_{\text{VII.6}} = \frac{K_{\beta\text{-}\alpha}}{K_{\alpha\text{-}\beta}} = \frac{x_{A\text{-}\beta} f_{A\text{-}\beta} x_{B\text{-}\alpha} f_{B\text{-}\alpha}}{x_{B\text{-}\beta} f_{B\text{-}\beta} x_{A\text{-}\alpha} f_{A\text{-}\alpha}} = \frac{a_{A\text{-}\beta} a_{B\text{-}\alpha}}{a_{B\text{-}\beta} a_{A\text{-}\alpha}}$$

where f is the partial activity coefficient and a the partial activity. The product of the f's appears as φ in Eq. (VII.7). The term "partial" is used to distinguish between the activity of A on the site from the activity of A in the crystal.

The distribution coefficient is

$$K_D = \frac{x_{A\text{-}\beta} x_{B\text{-}\alpha}}{x_{B\text{-}\beta} x_{A\text{-}\alpha}},$$

K_D has sometimes been referred to as the ordering parameter. The distribution coefficient, however, should not be confused with the ordering parameter S used to describe ordering in alloys. $S = 1$ corresponds to the highest possible order, and $S = 0$, to complete disorder. This is opposite in the case of K_D. Further K_D will be used to describe order-disorder in nonstoichiometric silicates forming complete crystalline solution series. In such silicates the formation of a fully ordered or disordered periodic structure is not possible. Even with the greatest tendency towards ordering, some of the excess atoms of one component must inevitably occupy sites belonging to the other, which leads to a lower order or disorder. The distribution coefficient is a function of T and the varying ratio of A to B in the crystal.

The equilibrium constant $K_{\text{VII.6}}$ is a function of P and T only. However, as the volume changes involved in the ion exchange are negligible, the dependence of $K_{\text{VII.6}}$ on P is ignored, and $K_{\text{VII.6}}$ is considered to be only temperature dependent.

The definition of the chemical potential of a cation on a site presents certain problems (MUELLER, GHOSE and SAXENA, 1970). One may write the following equations, as done by GROVER and ORVILLE (1969), for the chemical potential of a cation A on the sites α and β:

$$\mu_{A\text{-}\alpha} = \mu^0_{A\text{-}\alpha} + RT \ln a_{A\text{-}\alpha} \qquad \text{(VII.9)}$$

and

$$\mu_{A\text{-}\beta} = \mu^0_{A\text{-}\beta} + RT \ln a_{A\text{-}\beta}. \qquad \text{(VII.10)}$$

However, this involves defining two different chemical potentials for one species in a single homogeneous phase which is incongruent with classical thermodynamics.

To avoid this difficulty, BORGHESE (1967) regards A in site α as a distinct species from A in site β. This is somewhat analogous to speaking of the chemical potentials of O_2 and O_3 in a homogeneous gas phase. The idea of defining a new potential analogous to chemical potential called a site preference potential (GREENWOOD in GROVER and ORVILLE, 1971) could also be considered. This aspect is discussed more in Chapter VIII.

The standard site preference energy or the intracrystalline ion-exchange energy ΔG^0 for the exchange reaction (Eq. VII.6) is given by

$$\Delta G^0 = -RT \ln K_{VII.6} \qquad (VII.11)$$

where $K_{VII.6}$ is the equilibrium constant and is a function of T only, unlike K_D, which is a function of both T and composition.

3. Thermodynamic Functions of Mixing

One of the principal aims of the study of order-disorder phenomena is to investigate the thermodynamic properties of the crystalline solution as a whole. In case of an ideal macrophase, the activity-composition relation is given by

$$a_A = (x_A)^N \qquad (VII.12)$$

where N is the number of structural sites in the crystal. When there are two sites,

$$a_A = (x_A)^2$$

and if these sites are different,

$$a_A = (x_{A-\alpha} + x_{A-\beta}) \qquad (VII.13)$$

or

$$a_A = (x_{A-\alpha}) (x_{A-\beta}) \qquad (VII.14)$$

where α and β are two nonequivalent structural sites. The latter method has been generally used (MUELLER, 1962; THOMPSON, 1969). The difference between the two models is that in the latter individual sites are regarded as ideal solutions. The overall crystalline solution, therefore, cannot be ideal in this model and there would be a negative deviation from ideal solution (see Fig. 33). A comparison between the entropy of mixing corresponding to the two relations (VII.13) and (VII.14) may be made as follows. The entropy of mixing for the crystalline solution as a whole is given by:

$$S^M = -R(x_A \ln x_A + x_B \ln x_B).$$

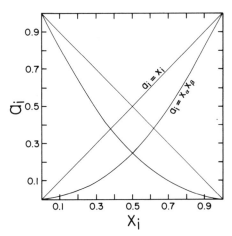

Fig. 33. Activity-composition relation expressed in two ways for a disordered crystal. $a_i = x_i$ expresses that the crystalline solution as a whole is ideal while $a_i = x_\alpha x_\beta$ represents the ideal mixing on sites α and β

The expression for entropy according to Eq.(VII.13) on one cation basis is

$$S^M = -R\left(\frac{x_{A\text{-}\alpha} + x_{A\text{-}\beta}}{2} \ln \frac{x_{A\text{-}\alpha} + x_{A\text{-}\beta}}{2} + \frac{x_{B\text{-}\alpha} + x_{B\text{-}\beta}}{2} \right.$$
$$\left. \cdot \ln \frac{x_{B\text{-}\alpha} + x_{B\text{-}\beta}}{2}\right) \tag{VII.15}$$

while according to (VII.14) we have

$$S^M = -R\left[\frac{x_{A\text{-}\alpha} + x_{A\text{-}\beta}}{2} \ln\{(x_{A\text{-}\alpha})(x_{A\text{-}\beta})\}^{\frac{1}{2}} + \frac{x_{B\text{-}\alpha} + x_{B\text{-}\beta}}{2} \right.$$
$$\left. \cdot \ln\{(x_{B\text{-}\alpha})(x_{B\text{-}\beta})\}^{\frac{1}{2}}\right] \tag{VII.16}$$

Eq.(VII.16) may be rearranged as

$$S^M = -R/4\{(x_{A\text{-}\alpha} \ln x_{A\text{-}\alpha} + x_{A\text{-}\beta} \ln x_{A\text{-}\beta})$$
$$+ (x_{B\text{-}\alpha} \ln x_{B\text{-}\alpha} + x_{B\text{-}\beta} \ln x_{B\text{-}\beta})\} + \text{cross terms} \tag{VII.17}$$
$$= -R/4\{S_\alpha^M + S_\beta^M\} + \text{cross terms}$$

where S_α^M and S_β^M are the ideal entropy of mixing of A and B on the sites. It may be shown by computation that

$$S^M(\text{Eq.(VII.16)}) > S^M(\text{Eq.(VII.15)}) \tag{VII.18}$$

which would mean negative deviation from ideal solution and an increased stability for the solution. The greater entropy value from Eq. (VII.16) is evidently due to the fact that there can be a greater disorder in solutions when there are two or more structural sites. For the non-ideal solution we have,

$$a_i = (a_{i-\alpha})(a_{i-\beta}) \tag{VII.19}$$

where $a_{i-\alpha}$ and $a_{i-\beta}$ are partial activities referring to the sites. In an orthopyroxene ($MgMgSi_2O_6$–$FeFeSi_2O_6$) where there are two sites, M1 and M2, the activity of Fe^{2+} in the crystal may be expressed as

$$a_{Fe\text{-}opx} = a_{Fe\text{-}M1}\, a_{Fe\text{-}M2}. \tag{VII.20}$$

If the activity is considered on a one-cation basis, i.e., for the crystal ($MgSiO_3$–$FeSiO_3$),

$$a_{Fe\text{-}opx} = (a_{Fe\text{-}M1}\, a_{Fe\text{-}M2})^{\frac{1}{2}} \tag{VII.21}$$

The partial activity $a_{i-\alpha}$ is equal to $f_{i-\alpha} x_{i-\alpha}$ where f is the partial activity coefficient. At a certain temperature the atomic ratio x_i in the two sites α and β can be determined by X-ray or other resonance techniques. The next problem involves the evaluation of the partial activity coefficients.

Several crystals of suitable composition $(A, B)M$ between the end members AM and BM may be chosen and heated at a certain temperature for a time long enough to attain equilibrium for the intracrystalline ion-exchange (Eq. (VII.6)). Several such distribution isotherms may be obtained. A model suitable for interrelating the partial activity co-efficient with the atomic fraction at the site must be found. The simple mixture or regular solution model may be found useful in cases where the form of the distribution isotherms does not indicate too much of a nonideal state of mixing A and B at α and β. Thus,

$$\ln K_{VII.6} = \ln K_D - \frac{W_\alpha}{RT}(1 - 2x_{A\text{-}\alpha}) + \frac{W_\beta}{RT}(1 - 2x_{A\text{-}\beta}) \tag{VII.22}$$

where W is related to the partial activity coefficient by

$$RT \ln f_{A\text{-}\alpha} = W_\alpha(1 - x_{A\text{-}\alpha})^2$$

At this point, certain other partial functions of mixing may be considered. The partial free energy of mixing at the sites is given by

$$G^M = x_{A\text{-}\alpha} RT \ln a_{A\text{-}\alpha} + x_{B\text{-}\alpha} RT \ln a_{B\text{-}\alpha} \tag{VII.23}$$

$$G^M = x_{A\text{-}\alpha} RT \ln f_{A\text{-}\alpha} + x_{B\text{-}\alpha} RT \ln f_{B\text{-}\alpha}$$
$$+ T(x_{A\text{-}\alpha} R \ln x_{A\text{-}\alpha} + x_{B\text{-}\alpha} R \ln x_{B\text{-}\alpha}). \tag{VII.24}$$

Substituting $S_\alpha^M = -R(x_{A-\alpha} \ln x_{A-\alpha} + x_{B-\alpha} \ln x_{B-\alpha})$ into Eq. (VII.24), and from Eq. (II.5)

$$G_\alpha^M = W_\alpha(x_{A-\alpha} x_{B-\alpha}) - T S_\alpha^M. \tag{VII.25}$$

The term $W(x_{A-\alpha} x_{B-\alpha})$ is also the partial excess free energy of mixing.

In a crystalline solution such as AAM–BBM, the total free energy of mixing is given by

$$G^M = \frac{x_{A-\alpha} + x_{A-\beta}}{2} RT \ln a_{A-\alpha} a_{A-\beta} + \frac{x_{B-\alpha} + x_{B-\beta}}{2} RT \ln a_{B-\alpha} a_{B-\beta} \tag{VII.26}$$

which can be shown to be

$$G^M = \frac{x_{A-\alpha} - x_{A-\beta}}{2} RT \ln K_{VII.6} + G_\alpha^M + G_\beta^M, \tag{VII.27}$$

$$G^M = \frac{x_{A-\alpha} - x_{A-\beta}}{2} RT \ln K_{VII.6} - T(S_\alpha^M + S_\beta^M) + W_\alpha x_{B-\alpha} x_{A-\alpha} + W_\beta x_{A-\beta} x_{B-\beta}. \tag{VII.28}$$

The presence of the term $RT \ln K_{VII.6}$ in the expressions (VII.27) and (VII.28) imply that the free energy of mixing in the crystal as a whole can be calculated only for those crystals which have attained intracrystalline equilibrium at a given temperature. Note also the important point that the sum of the free energy of mixing on the individual site is not the free energy of mixing for the crystal.

Expression (VII.28) is similar to the one derived by GROVER and ORVILLE (1969) for ideal mixing at the sites. Note that in the above expression $K_{VII.6}$ is the equilibrium constant and not the distribution coefficient, as in the case of ideal mixing.

Substituting $\Delta G^0 = -RT \ln K_{VII.6}$ into Eq. (VII.28),

$$G^M = \frac{x_{A-\beta} - x_{A-\alpha}}{2} \Delta G^0 - T(S_\alpha^M + S_\beta^M) + (G_\alpha^{EM} + G_\beta^{EM}) \tag{VII.29}$$

Thus the free energy of mixing in the crystal as a whole is a result of the energy due to the distribution of the cation between α and β sites, the entropy change due to the distribution of A and B within α and β sites, and finally the excess energy of mixing that is the result of the nonideal state of the solution in α and β.

The partial excess free energies of mixing at the sites are

$$G_\alpha^{EM} = H_\alpha^{EM} - T S_\alpha^{EM}, \tag{VII.30}$$

$$G_\beta^{EM} = H_\beta^{EM} - T S_\beta^{EM}. \tag{VII.31}$$

Substituting Eqs. (VII.30) and (VII.31) into Eq. (VII.29),

$$G^M = \frac{x_{A-\beta} - x_{A-\alpha}}{2} \Delta G^0 + H_\alpha^{EM} + H_\beta^{EM} - T(S_\alpha^{IM} + S_\alpha^{EM} + S_\beta^{IM} + S_\beta^{EM}). \tag{VII.32}$$

The thermodynamic relations presented in this and earlier sections will be used in analyzing the data on site occupancies in orthopyroxene later.

4. Kinetics of Order-Disorder

The time rates of order-disorder are important for understanding the cooling history of the crystals. The thermometric aspect of order-disorder process is discussed later but presently, a brief review of the theory should be in order. MUELLER (1967, 1969) suggested that the time rate of change of A on sublattice α in a quasi-binary solution with n sites may be expressed by the equation

$$\frac{dx_i}{dt} = C_0 \sum_{j=1}^{n-1} V_{ji} K_{ji} \varphi_{ji} x_j (1 - x_i) - C_0 \sum_{j=1}^{n-1} V_{ij} K_{ij} \varphi_{ij} x_i (1 - x_j) \qquad \text{(VII.33)}$$

where C_0 is the total concentration of all sites, x is the site occupancy and V is the stoichiometric coefficient relating the concentrations on individual sites to the total concentration C_0. K's and φ's are as in Eq. (VII.7). When nonequivalent sites have the same number, V_{ij} is equal to V_{ji}. For crystals with two sites α and β and assuming $\varphi_{\alpha\text{-}\beta} = \varphi_{\beta\text{-}\alpha} = 1$ (see MUELLER, 1967), we may simplify Eq. (VII.33) for a quasi-binary crystal $(A, B) M$ as

$$-\frac{dx_\alpha}{dt} = \tfrac{1}{2} C_0 K_{\alpha\text{-}\beta} [(1 - K_{\beta\text{-}\alpha}^0)(x_\alpha)^2 + (2K_{\beta\text{-}\alpha}^0 x - 2x + K_{\beta\text{-}\alpha}^0 + 1)x - 2K_{\beta\text{-}\alpha}^0 x]$$

$$\text{(VII.34)}$$

where

$$K_{\beta\text{-}\alpha}^0 = \frac{K_{\beta\text{-}\alpha}}{K_{\alpha\text{-}\beta}} . \qquad \text{(VII.35)}$$

x is the mole fraction in the crystal as a whole and is equal to

$$x = \tfrac{1}{2} x_\alpha + \tfrac{1}{2} x_\beta \qquad \text{(VII.36)}$$

where x_α and x_β are site occupancies. At equilibrium since $dx_\alpha/dt = 0$, we have

$$K_{\beta\text{-}\alpha}^0 = \frac{K_{\beta\text{-}\alpha}}{K_{\alpha\text{-}\beta}} = \frac{x_\alpha(1 - x_\beta)}{x_\beta(1 - x_\alpha)} . \qquad \text{(VII.37)}$$

Substituting for x from Eq. (VII.36) in Eq. (VII.37), we obtain

$$K_{\beta\text{-}\alpha}^0 = \frac{x_\alpha(1 - 2x + x_\alpha)}{(2x - x_\alpha)(1 - x_\alpha)} . \qquad \text{(VII.38)}$$

Eq. (VII.34) may be integrated directly for a fixed bulk composition x and for any temperature at which $K_{\alpha-\beta}$ and $K^0_{\beta-\alpha}$ are defined. The expression is as follows:

$$\mp \tfrac{1}{2} C_0 K_{\alpha-\beta} \Delta t = \frac{1}{2m^{\frac{1}{2}}} \ln \frac{n \mp m^{\frac{1}{2}}}{\pm n + m^{\frac{1}{2}}} \Big|^{x''}_{x'} \qquad \text{(VII.39)}$$

where the upper signs represent equation for isothermal ordering of the crystal and the lower signs for isothermal disordering, Δt is the time interval and x'_α and x''_α are the values of x_α at the upper and lower limit of integration and m and n are given by

$$m = \tfrac{1}{4}(2 K^0_{\beta-\alpha} x - 2x + K^0_{\beta-\alpha} + 1)^2 + (1 - K^0_{\beta-\alpha})(2 K^0_{\beta-\alpha} x), \qquad \text{(VII.40)}$$

$$n = (1 - K^0_{\beta-\alpha}) x + \tfrac{1}{2}(2 K^0_{\beta-\alpha} x - 2x + K^0_{\beta-\alpha} + 1). \qquad \text{(VII.41)}$$

The integrand for ordering is defined for the range $n^2 > m$ and for disordering for the range $n^2 < m$ and it has an infinite discontinuity at values of x_α, x and $K^0_{\beta-\alpha}$.

If the data on equilibrium constant $K^0_{\beta-\alpha}$ and the site occupancy x_α as a function of Δt is provided by experiments, we may determine the rate constant. Such data are available on one orthopyroxene ($x_{Mg} = 0.574$) at 500, 600, 700, and 1000° C and Fig. 7 in VIRGO and HAFNER (1969) shows a plot of $K_{21} \Delta t$ against x_{Mg-M2} in orthopyroxene. For the case of disordering K_{21} would correspond to K_{M2-M1} in the present notation. These data yield approximate rate constants (K_{M2-M1}) as 6×10^{-5}/min at 500 and 1×10^{-2}/min at 1000° C (VIRGO and HAFNER, 1969). The activation energy (defined as the change in free energy in going from the reactant to the activated state) in the direction of disordering ($= -RT \ln K_{M2-M1}$) is approximately 20 Kcal. As compared to order-disorder in some other systems such as Al–Si order-disorder, the rate constants for Fe–Mg exchange are high and the activation energy for disordering (or ordering) is low. These data are important in understanding the thermal history of the rocks as discussed by MUELLER (1969, 1970) and in Chapter XI.

VIII. Pyroxene Crystalline Solution

1. Orthopyroxene

Orthopyroxene is one of the few important rock-forming minerals that can be considered as quasi-binary without significant loss of accuracy. Usually more than 95% of the mineral is a crystalline solution of the end members enstatite ($MgSiO_3$) and ferrosilite ($FeSiO_3$). Fe^{2+} and Mg^{2+} are distributed between two nonequivalent sites M 1 and M 2. With the use of X-ray or Mössbauer spectroscopic technique, it is possible to determine the proportion of Fe^{2+} in the two nonequivalent sites (EVANS, GHOSE and HAFNER, 1967). These data can be used with the help of suitable solution models to determine the thermodynamic properties of the solution (VIRGO and HAFNER, 1969; SAXENA and GHOSE, 1971).

a) Intersite Ion Exchange

The thermodynamics of the intracrystalline ion exchange was discussed in Chapter VII. M 1 and M 2 sites may be regarded as two interpenetrating subsystems, each with its own thermodynamic properties of mixing. Analogous to heterogeneous ion-exchange equilibria, the ion-exchange reaction may be written

$$Fe^{2+}(M\,2) + Mg^{2+}(M\,1) \rightleftharpoons Fe^{2+}(M\,1) + Mg^{2+}(M\,2) \qquad (VIII.1)$$

The equilibrium constant for the above reaction at a certain P and T is

$$K_{VIII.1} = \frac{x_{Fe\text{-}M1}\, f_{Fe\text{-}M1}\, x_{Mg\text{-}M2}\, f_{Mg\text{-}M2}}{x_{Fe\text{-}M2}\, f_{Fe\text{-}M2}\, x_{Mg\text{-}M1}\, f_{Mg\text{-}M1}}. \qquad (VIII.2)$$

$K_{VIII.1}$ is mainly a function of T. P has little influence. The standard free energy for ion exchange at a certain T is

$$\Delta G^0_{VIII.1} = -RT \ln K_{VIII.1}. \qquad (VIII.3)$$

This energy is part of the total Gibbs free energy of the crystal and is, therefore, an important thermodynamic quantity. To determine $\Delta G^0_{VIII.1}$ or $K_{VIII.1}$, a determination of the atomic fractions $x_{Fe\text{-}M1}$ and $x_{Fe\text{-}M2}$ and a determination of the partial activity coefficients (f values), which are functions of T and composition, must be made. While the atomic fractions or the site occupancies can be determined quantitatively by using X-ray or Mössbauer techniques, the partial activity coefficients cannot be determined without the use of a solution model.

b) Order-Disorder on Individual Sites and the Choice of a Solution Model

Figure 34 (GHOSE, 1965, Figs. 1 and 2) shows a scheme of ordering in the orthopyroxene crystal as a whole. In hypersthene ($Fs_{50}En_{50}$) there is the possibility of complete occupation of M 1 sites by Mg^{2+} and of M 2 sites by Fe^{2+}. For all other compositions, one of the two ions must inevitably occupy sites other than they would normally. In fact, as suggested before, it is to be expected that the two ions will always show some kind of equilibrium distribution over the two sites as a function of T and composition. Although hypersthene has the right composition to be completely ordered, the kinetics of the ion exchange below a certain transition temperature precludes the ordered structure shown in Fig. 34.

To consider the ion exchange in Eq. (VIII.1) M 1 and M 2 must be considered individually as subsystems. Complete order in the crystal as a whole as shown in Fig. 34 also means complete order on the sites themselves. For all other compositions, should be considered whether two neighboring M 1 or M 2 sites are both occupied by the same cation or by two different cations. It may be assumed that the occupancy of M 1

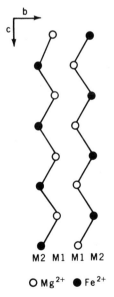

M2 M1 M1 M2

O Mg^{2+} ● Fe^{2+}

Fig. 34. Mg^{2+}-Fe^{2+} ordering scheme in an orthopyroxene (GHOSE, 1965)

and M 2 sites by Mg^{2+} and Fe^{2+} is disordered and that the solutions at the two sites approximate the simple mixture model. The effect of ordering on the sites by using the quasi-chemical approximation will be considered later, and the results of these two models will be compared.

c) Determination of Site Occupancy in Heated Orthopyroxenes

By using Mössbauer spectroscopy, VIRGO and HAFNER (1969) determined a distribution isotherm at $1273\,°K$. SAXENA and GHOSE (1971) similarly determined isotherms between 773 and $1073\,°K$ (Fig. 35).

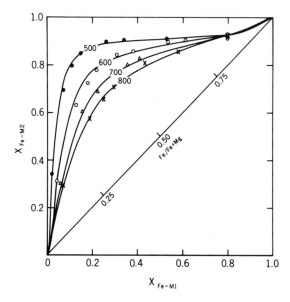

Fig. 35. Distribution isotherms for the Mg^{2+}- Fe^{2+} partitioning between M1 and M2 sites in heated natural orthopyroxenes (SAXENA and GHOSE, 1971). The curves are least squares fit to the experimental data using simple mixture model for the sites

Orthopyroxene crystals are heated for a length of time sufficient to bring about the ion-exchange equilibrium. The samples are then quenched, and the site occupancy is determined. Note that the site occupancy is averaged over all the sites present in the crystals and is a function of the changes that take place not only in the neighboring site but also in the remaining Si–O structural framework.

d) Sites as Simple Mixtures

The distribution isotherms can be represented by

$$\ln K_{\text{VIII.1}} = \ln K_D + \frac{W_{M1}}{RT}(1 - 2x_{\text{Fe-M1}}) - \frac{W_{M2}}{RT}(1 - 2x_{\text{Fe-M2}})$$

where

$$K_D = x_{\text{Fe-M1}}(1 - x_{\text{Fe-M2}})/x_{\text{Fe-M2}}(1 - x_{\text{Fe-M1}})$$

x_{Fe} is the site occupancy, and W is the energy constant of the simple mixture model. A nonlinear least-squares program is used to calculate $K_{\text{VIII.1}}$, W_{M1}, and W_{M2} by simultaneous iterations on all the three unknown variables. Fig. 36 shows a plot of $\Delta G^0_{\text{VIII.1}}$ (the standard free energy of ion exchange $= -RT \ln K_{\text{VIII.1}}$), W_{M1}, and W_{M2} against $1/T$. There is a linear relation between $1/T$ and all three variables at 873, 973, and 1073 °K. The equations to the three straight lines are

$$W_{M1} = 3525\left(\frac{10^3}{T}\right) - 1667, \qquad \text{(VIII.4)}$$

$$W_{M2} = 2458\left(\frac{10^3}{T}\right) - 1261, \qquad \text{(VIII.5)}$$

$$\Delta G^0_{\text{VIII.1}} = 4479 - 1948\left(\frac{10^3}{T}\right). \qquad \text{(VIII.6)}$$

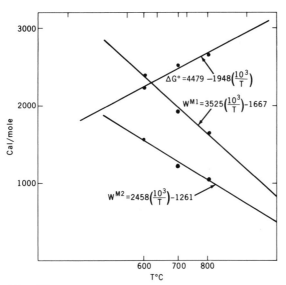

Fig. 36. W^{M1}, W^{M2} and ΔG° plotted against $1/T$. W's are the energy constant of the simple mixture model and ΔG° is the standard free energy of intersite ion-exchange

The values at 773 °K, however, do not fall on the straight lines. This may be due to a low degree of accuracy in determining the M 1 site occupancy, particularly in Mg-rich samples at 773 °K. The constants at 1273 °K may be calculated by extrapolation. An isotherm at this temperature plotted with these constants is consistent for the most part with the data of VIRGO and HAFNER (1969).

e) Thermodynamic Functions of Mixing with Sites as Simple Mixtures

The partial functions of mixing in M 1 and M 2 may be calculated from the relations

$$G_{M1}^{EM} = x_{Fe\text{-}M1} \, x_{Mg\text{-}M1} \, W_{M1}$$

$$- S_{M1}^{EM} = x_{Fe\text{-}M1} \, x_{Mg\text{-}M1} \, \frac{\partial W_{M1}}{\partial T}$$

and

$$H_{M1}^{EM} = x_{Fe\text{-}M1} \, x_{Mg\text{-}M1} \left(W_{M1} - T \frac{\partial W_{M1}}{\partial T} \right).$$

The M 2 site relations are similarly calculated.

G^{EM} for the crystal as a whole is given by

$$G_{opx}^{EM} = R T (x_{Fe\text{-}opx} \ln f_{Fe\text{-}opx} + x_{Mg\text{-}opx} \ln f_{Mg\text{-}opx}). \qquad \text{(VIII.7)}$$

The values of the activity coefficients f may be obtained by

$$f_{Fe\text{-}opx} = \frac{\left[x_{Fe\text{-}M1} \exp\frac{W_{M1}}{RT}(1 - x_{Fe\text{-}M1})^2 \right]^{1/2} \left[x_{Fe\text{-}M2} \exp\frac{W_{M2}}{RT}(1 - x_{Fe\text{-}M2})^2 \right]^{1/2}}{x_{Fe\text{-}opx}}$$

$$\text{(VIII.8)}$$

$f_{Mg\text{-}opx}$ is similarly calculated.

It is useful to represent G^{EM} as a polynomial in x as suggested by GUGGENHEIM (1937):

$$G_{opx}^{EM} = x_{Fe\text{-}opx} x_{Mg\text{-}opx} \left[A_0 + A_1 (x_{Fe\text{-}opx} - x_{Mg\text{-}opx}) + A_2 (x_{Fe\text{-}opx} - x_{Mg\text{-}opx})^2 + \cdots \right]$$

which may also be written as

$$G_{opx}^{EM}/x_{Fe\text{-}opx} x_{Mg\text{-}opx} = A_0 + A_1 (x_{Fe\text{-}opx} x_{Mg\text{-}opx}) + A_2 (x_{Fe\text{-}opx} - x_{Mg\text{-}opx})^2 + \cdots$$

$$\text{(VIII.9)}$$

Eq. (VIII.9) is of the form

$$Y = a_0 + a_1 x_1 + a_2 x_1^2 + \cdots$$

and may be solved by a least-squares program. To determine the constants A_0, A_1, and A_2, G_{opx}^{EM} is obtained from Eqs. (VIII.7) and (VIII.8). The

Table 8. Values of equilibrium constant $K_{VIII.1}$, $\Delta G^0_{VIII.1}$, and the constants W_{M1} and W_{M2}

$T^\circ(K)$	$K_{VIII.1}$	W_{M1} cal/mole	W_{M2} cal/mole	$\Delta G^0_{VIII.1}$ cal/mole
773	0.279	2893	1919	1959
873	0.277	2390	1577	2233
973	0.273	1916	1215	2509
1073	0.289	1641	1057	2646
1173	0.298	1338	834	2818
1273	0.311	1102	670	2949

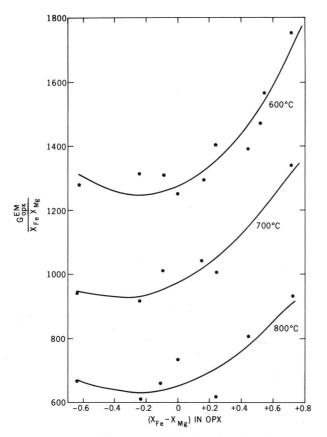

Fig. 37. Reduced excess free energy of mixing plotted against $(x_{Fe} - x_{Mg})$ in ortho-pyroxene. The curves are least-squares fits using the equation (VIII.9). This fit yields the constants A_0, A_1 and A_2

constants listed in Table 8 are used in Eq. (VIII.8). These values of excess free energy are then used to calculate the three constants A_0, A_1, and A_2 shown in Table 9 by a least-squares fit to Eq. (VIII.9). Fig. 37 shows three curves at 873, 973, and 1073 °K fitted to the data on reduced excess free energy $(G^{EM}/x_{Fe}x_{Mg})$ against $(x_{Fe} - x_{Mg})$. This method of determining A_0, A_1, and A_2 is more reliable than the Redlich and Kister method used by SAXENA and GHOSE (1971).

Table 9. The three constants A_0, A_1, and A_2 obtained by a least-square fit to Eq. (VIII.9). The values in parentheses are values calculated by using Eqs. (VII.10) to (VII.12)

$T°(K)$	A_0 cal/mole	A_1 cal/mole	A_2 cal/mole
773	1544	478	867
873	1278	287	461
973	951	199	335
1073	693	168	260
1173	529	170	163

The three constants as a function of $1/T$ (Fig. 38) are given by

$$A_0 = 10802 - 33862 \left(\frac{10^3}{T}\right) + 35135 \left(\frac{10^3}{T}\right)^2 - 11202 \left(\frac{10^3}{T}\right)^3, \quad \text{(VIII.10)}$$

$$A_1 = 1789 - 3612 \left(\frac{10^3}{T}\right) + 2008 \left(\frac{10^3}{T}\right)^2, \quad \text{(VIII.11)}$$

and

$$A_2 = -13863 + 41051 \left(\frac{10^3}{T}\right) - 40299 \left(\frac{10^3}{T}\right)^2 + 13426 \left(\frac{10^3}{T}\right)^3. \quad \text{(VIII.12)}$$

The heat of mixing and the entropy of mixing can now be readily calculated with the following equations:

$$-S_{opx}^{EM} = x_{Fe\text{-}opx} x_{Mg\text{-}opx} \left[\frac{\partial A_0}{\partial T} + \frac{\partial A_1}{\partial T} (x_{Fe\text{-}opx} - x_{Mg\text{-}opx}) \right.$$
$$\left. + \frac{\partial A_2}{\partial T} (x_{Fe\text{-}opx} - x_{Mg\text{-}opx})^2 + \cdots \right] \quad \text{(VIII.13)}$$

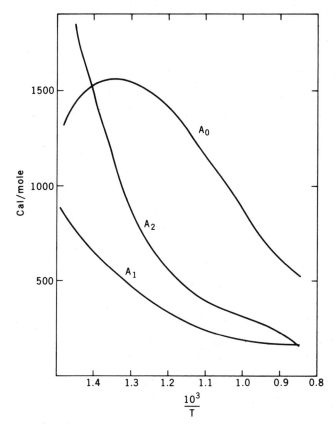

Fig. 38. Least squares curves showing A_0, A_1 and A_2 as a function $1/T$ for orthopyroxene. The equations to the curves are (VIII.10) (VIII.11) and (VIII.12)

and

$$H_{opx}^{EM} = x_{Fe\text{-}opx} x_{Mg\text{-}opx} \left\{ A_0 - T\left(\frac{\partial A_0}{\partial T}\right) + \left[A_1 - T\left(\frac{\partial A_1}{\partial T}\right)\right](x_{Fe\text{-}opx} - x_{Mg\text{-}opx}) \right.$$

$$\left. + \left[A_2 - T\left(\frac{\partial A_2}{\partial T}\right)\right](x_{Fe\text{-}opx} - x_{Mg\text{-}opx})^2 + \cdots \right\}. \qquad (VIII.14)$$

The free energy of mixing is given by

$$G_{opx}^{M} = H_{opx}^{EM} - T S_{opx}^{EM} + R T(x_{Fe\text{-}opx} \ln x_{Fe\text{-}opx} + x_{Mg\text{-}opx} \ln x_{Mg\text{-}opx}). \qquad (VIII.15)$$

Activity-composition relations for the orthopyroxene crystal as a whole at 600 and 800° C are shown in Figs. 39 and 40.

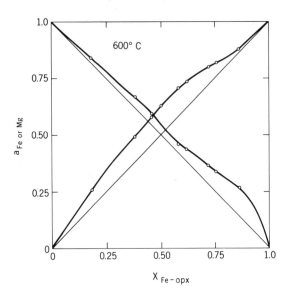

Fig. 39. Activity-composition relation in orthopyroxene at 600° C with sites as simple mixtures

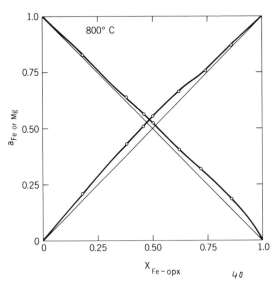

Fig. 40. Activity-composition relation in orthopyroxene at 800° C with sites as simple mixtures

It may be useful to write an expression for the free energy of mixing of the crystal as a whole in terms of partial thermodynamic functions of mixing referring to the sites. For $(Mg, Fe)_2Si_2O_6$, according to Eq.(VII.23)

$$G^M_{opx} = \frac{x_{Fe-M1} - x_{Fe-M2}}{2} \Delta G^0_{VIII.1} + H^{EM}_{M1} + H^{EM}_{M2} - T(S^{IM}_{M2} + S^{IM}_{M2} + S^{EM}_{M1} + S^{EM}_{M2}).$$

(VIII.16)

Note that $H^{EM}_{opx} \neq H^{EM}_{M1} + H^{EM}_{M2}$ and similarly $S^{EM}_{opx} \neq S^{EM}_{M1} + S^{EM}_{M2}$. Eq.(VIII.16) divides the total free energy of mixing into three terms. The term in the first bracket is the potential from the difference in site occupancy energy. It includes both the contributions of the standard enthalpy and entropy of the exchange. The terms in the second and third brackets include the heat of mixing and the entropy of mixing, respectively, at the individual sites.

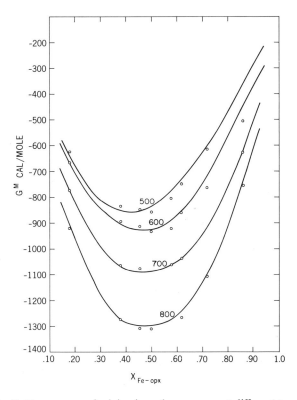

Fig. 41. Free energy of mixing in orthopyroxene at different temperatures

This division of the free energy of mixing is artificial. It may be useful in understanding the change in free energy as a function of temperature and order-disorder. Figs. 41, 42 and 43 show a comparison

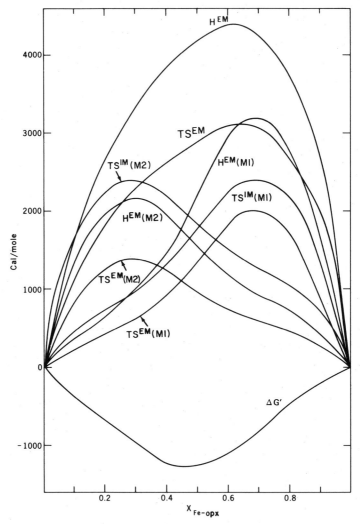

Fig. 42. Thermodynamic functions of mixing in orthopyroxene at 600° C, both at the sites and in the crystal as a whole plotted against $x_{Fe\text{-}opx}$. The functions of mixing at the sites correspond to the atomic fractions at M1 and M2, which in turn correspond to an equilibrium distribution at 600° C in the crystal with $x_{Fe\text{-}opx}$ as measured on the abscissa. $\Delta G'$ is

$$\{(x_{Fe\text{-}M1} - x_{Fe\text{-}M2})/2\} \Delta G^0_{VIII.1}$$

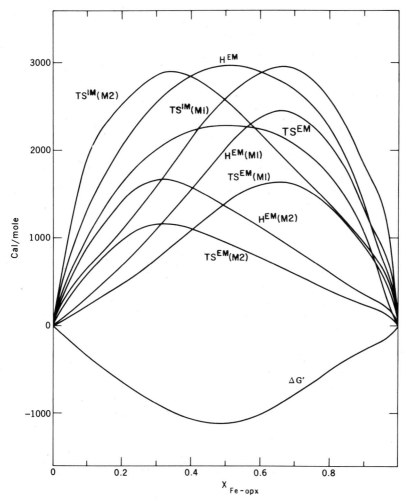

Fig. 43. Thermodynamic functions of mixing in orthopyroxene at 800° C plotted similarly as in Fig. 42

of the various functions of mixing at various temperatures. The values of H^{EM} and TS^{EM} for the crystal as a whole are obtained from Eq. (VIII.13) and (VIII.14). It may be noted that the first energy term

$$\frac{x_{\text{Fe-M1}} - x_{\text{Fe-M2}}}{2} \, \Delta G^0_{\text{VIII.1}}$$

along with the entropy terms increase the negative G^M_{opx} and make the solution stable. The negative value of the first energy term decreases somewhat with increasing temperature, indicating as if the disordering works against stability of the solution. This effect is, however, limited and overridden by the positive entropy of mixing at the sites.

At 1073 °K, the contributions of the partial energies of mixing at the sites to the energies of mixing in the crystal as a whole are somewhat asymmetric. The mixing energies H_{opx}^{EM} and TS_{opx}^{EM} have more or less symmetric values. With decreasing temperature, the energies of mixing at the sites increase, those for the M 1 site increasing relatively more than those for the M 2 site. The H_{opx}^{EM} and S_{opx}^{EM}, therefore, become more and more asymmetric.

According to a preliminary observation by VIRGO and HAFNER (1969), the M 1 polyhedra is a little distorted at the enstatite end and becomes more and more regular with increasing Fe/Mg ratio. The M 2 polyhedra is quite distorted at the enstatite end and the distortion increases with increasing Fe/Mg ratio. These changes in the polyhedra are directly correlated with the entropy and enthalpy of mixing on the sites. As the regularity increases in M 1, the energies of mixing increase. As the distortion increases in M 2, the energies of mixing decrease. Thus we require to put more energy into the system to increase the regularity of M 1. Energy requirements become less as distortion increases in M 2. These results indicate that it may be possible to predict the changes in the M 1 and M 2 polyhedral spaces as a function of temperature and composition by determining functions of mixing at the sites.

f) Sites as Solutions with Quasi-chemical Approximation

As shown in Chapter VI, the relation among $K_{VIII.1}$, W_{M1}, W_{M2}, and the site occupancies may be expressed by

$$K_D \frac{\left[1 + \dfrac{\varphi_{Mg\text{-}M1}(\beta - 1)}{\varphi_{Fe\text{-}M1}(\beta + 1)}\right]^{z\,q_{Fe\text{-}M1}/2} \left[1 + \dfrac{\varphi_{Fe\text{-}M2}(\beta' - 1)}{\varphi_{Mg\text{-}M2}(\beta' + 1)}\right]^{z'\,q_{Mg\text{-}M2}/2}}{\left[1 + \dfrac{\varphi_{Fe\text{-}M1}(\beta - 1)}{\varphi_{Mg\text{-}M1}(\beta + 1)}\right]^{z\,q_{Mg\text{-}M1}/2} \left[1 + \dfrac{\varphi_{Mg\text{-}M2}(\beta' - 1)}{\varphi_{Fe\text{-}M2}(\beta' + 1)}\right]^{z'\,q_{Fe\text{-}M2}/2}} = K_{VIII.1}$$

$$(VIII.17)$$

where β and β' are for M 1 and M 2 respectively, and are given by

$$\beta = \left[1 + 4\varphi_{Fe\text{-}M1}\,\varphi_{Mg\text{-}M1}\left(e^{\frac{2W_{M1}}{zRT}} - 1\right)\right]^{1/2}$$

and

$$\beta' = \left[1 + 4\varphi_{Fe\text{-}M2}\,\varphi_{Mg\text{-}M2}\left(e^{\frac{2W_{M2}}{zRT}} - 1\right)\right]^{1/2}.$$

The φ's are defined as before in Eq. (I.37); z and z' are the number of neighboring cations in M 1 and M 2, respectively; and the q's are contact factors.

As the silicate framework does not significantly change its character as a function of temperature or Fe^{2+}/Mg^{2+} ratio, the sites M 1 and M 2 have definite configurations and the polyhedral geometry changes only somewhat with changing Fe^{2+}/Mg^{2+} ratio. It may, therefore, be assumed that the number of sites that are neighbors to any one M 1 site are two or four other M 1 sites and $q_{Mg-M1} = q_{Fe-M1} = 1$. A reference to Fig. 34 shows that two of the four M 1 sites are somewhat nearer to a central M 1. It may be noted that the two inner strips with M 1 sites and two outer strips with M 2 sites lie more or less in a plane. Because the two M 2 strips are separated by the intervening M 1 sites, it is only realistic to consider that the number of neighboring M 2 sites to any one M 2 site is only two. It may also be assumed that $q_{Fe-M2} = q_{Mg-M2} = 1$; and, in fact, $q = 1$ for all q and $z = z' = 2$ may be substituted into Eq. (VIII.17).

Table 10. Quasi-chemical parameters W_{M1}/RT, W_{M2}/RT, and $K_{VIII.1}$ as calculated by Eq. (VIII.17)

$T°(K)$	$K_{VIII.1}$	W_{M1}/RT	W_{M2}/RT	χ^2	
				QC	Simple mixture
873	0.293	1.71	1.26	0.027	0.017
1073	0.273	0.733	0.469	0.007	0.004

$$\chi^2 = \Sigma \quad \frac{[K(\text{calculated}) - K(\text{by least squares})]^2}{K(\text{by least squares})}.$$

The intracrystalline distribution data presented in Fig. 35 were used to determine the quasi-chemical parameters W_{M1}, W_{M2}, and the equilibrium constant $K_{VIII.1}$ by a least-squares analysis of Eq. (VIII.17). The distribution data at 773 °K were not used, and the number of data points at 973 °K was not found sufficient for a satisfactory convergence. The values of W_{M1}, W_{M2}, and $K_{VIII.1}$ at 873 and 1073 °K are listed in Table 10. Chi-square values for the simple mixture model and for the quasi-chemical approximation are not significantly different.

g) Activity-Composition Relation at 873 °K

The partial activity coefficients at the M 1 site may be calculated by

$$f_{Fe-M1} = \left[1 + \frac{\varphi_{Mg-M1}(\beta - 1)}{\varphi_{Fe-M1}(\beta + 1)}\right]^{z\,q_{Fe-M1}/2},$$

$f_{\text{Fe-M }2}$ is similarly calculated. The activity-composition relation in the crystal as a whole may be determined from

$$a_{\text{Fe-opx}} = (x_{\text{Fe-M }1}\, f_{\text{Fe-M }1})^{1/2}\, (x_{\text{Fe-M }2}\, f_{\text{Fe-M }2})^{1/2}\,.$$

The results of these calculations are presented in Fig. 44. The figure shows that there is very little difference between the activities calculated by assuming sites as simple mixtures or as solutions with quasi-chemical approximation.

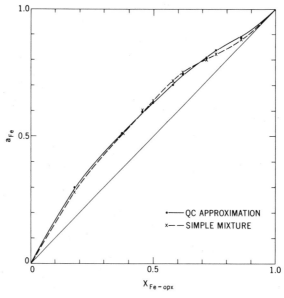

Fig. 44. Activity composition relation for $FeSiO_3$ in orthopyroxene crystalline solution at $600°$ C using simple mixture and quasi-chemical models for the sites

h) Unmixing

Although no data on unmixed orthopyroxenes are available, a possible unmixing can be predicted theoretically. Data on actual unmixing may not be available because of kinetic reasons. However, orthopyroxenes of suitable compositions, heated at temperatures close to the critical temperature of unmixing, may show certain structural characters, such as domains, by the use of electron microscopy.

The conditions of critical mixing are

$$\partial^2 G^{\text{EM}}/\partial x^2 = -RT/x(1-x)$$

and

$$\partial^3 G^{\text{EM}}/\partial x^3 = -RT(2x-1)/x^2(1-x)^2$$

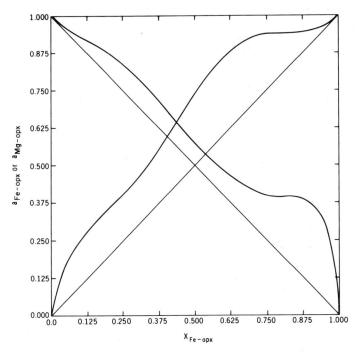

Fig. 45. Activity-composition relation in orthopyroxene at 450° C which is close to the temperature of unmixing; the critical composition is estimated to be $x_{Fe} = 0.83$

Successive differentiation of Eq. (VIII.9) leads to equations that are transcendental in character and cannot be solved without an iteration procedure. For the present purpose, it is enough to find an approximate temperature at which the activity-composition diagram shows the same activity over a small composition interval (Fig. 45). From Fig. 45, the critical temperature is in the vicinity of 723 °K and the critical composition is about $x_{Fe\text{-opx}} = 0.83$.

At 723 °K, the values of W_{M1}/RT and W_{M2}/RT are 2.23 and 1.49, respectively. For $x_{Fe\text{-opx}} = 0.83$, $x_{Fe\text{-}M1} = 0.75$ and $x_{Fe\text{-}M2} = 0.92$.

2. Calcic Pyroxenes

The thermodynamic functions of mixing in Ca-pyroxenes are more difficult to evaluate than those of orthopyroxenes for several reasons. First, unlike orthopyroxene, Ca-pyroxenes are ternary solutions in the system $CaSiO_3$–$MgSiO_3$–$FeSiO_3$. Second, within this ternary system, there is a region of immiscibility with structural transformations across the binodal as discussed in Chapter VI. Third, as shown below, to evaluate the activity-composition from data on site occupancies of Fe^{2+} and Mg, much more data are needed than in the case of orthopyroxene. The results presented in this section, therefore, are not as complete as those on the orthopyroxenes. Ca-pyroxenes considered here include augites, sub-calcic augites and pigeonites.

a) Intracrystalline Distribution of Fe^{2+}, Mg and Ca

As in orthopyroxene, Ca-pyroxenes have two nonequivalent sites M 1 and M 2. The coordination for M 1 is six but for M 2 it changes from six to eight when the site is occupied by Ca. X-ray crystal structural data (for example CLARK, APPLEMAN, and PAPIKE, 1969) indicate that almost all Ca in pyroxenes is in the M 2 site. Fe^{2+} and Mg are distributed between M 1 site and the remaining M 2 site. It may, therefore, be considered that M 1 site is a binary submixture of Fe and Mg components and M 2 site is a ternary submixture of Fe, Mg and Ca components. The intracrystalline ion exchange may be expressed as

$$Fe(M\,2) + Mg(M\,1) \rightleftharpoons Fe(M\,1) + Mg(M\,2) \qquad (VIII.18)$$

The equilibrium constant $K_{VIII.18}$ is

$$K_{VIII.18} = \left(\frac{x_{Fe\text{-}M1}\,x_{Mg\text{-}M2}}{x_{Mg\text{-}M1}\,x_{Fe\text{-}M2}}\right)\left(\frac{f_{Fe\text{-}M1}\,f_{Mg\text{-}M2}}{f_{Mg\text{-}M1}\,f_{Fe\text{-}M2}}\right) \qquad (VIII.19)$$

where as before x's are site occupancies and f's partial activity coefficients referring to activity at the sites. The partial activity coefficients are functions of T and the concentration of Mg and Fe and for M 2 also of the concentration of Ca. Eq. (VIII.19) may be written as:

$$\ln K_{D(VIII.18)} = \ln K_{VIII.18} - \ln\left(\frac{f_{Fe\text{-}M1}}{f_{Mg\text{-}M1}}\right) - \ln\left(\frac{f_{Mg\,M2}}{f_{Fe\,M2}}\right). \qquad (VIII.20)$$

A two constant equation for M1 site yields

$$\ln\left(\frac{f_{Fe\text{-}M1}}{f_{Mg\text{-}M1}}\right) = \frac{A_{0\text{-}M1}}{RT}(x_{Mg\text{-}M1} - x_{Fe\text{-}M1}) + \frac{A_{1\text{-}M1}}{RT}(6x_{Mg\text{-}M1}\,x_{Fe\text{-}M1} - 1).$$

$$(VIII.21)$$

Using Wohl's expression (II.40) for the ternary M 2, we have

$$\ln f_{\text{Mg-M 2}} = E_{\text{Mg-Fe}}(x_{\text{Fe}}^2 - 2x_{\text{Mg}}x_{\text{Fe}}^2 + 1/2x_{\text{Fe}}x_{\text{Ca}} - x_{\text{Mg}}x_{\text{Fe}}x_{\text{Ca}})$$
$$+ E_{\text{Fe-Mg}}(2x_{\text{Mg}}x_{\text{Fe}}^2 + 1/2x_{\text{Fe}}x_{\text{Ca}} + x_{\text{Mg}}x_{\text{Fe}}x_{\text{Ca}}) + E_{\text{Mg-Ca}}(x_{\text{Ca}}^2$$
$$- 2x_{\text{Mg}}x_{\text{Ca}}^2 + 1/2x_{\text{Fe}}x_{\text{Ca}} - x_{\text{Mg}}x_{\text{Fe}}x_{\text{Ca}}) + E_{\text{Ca-Mg}}(2x_{\text{Mg}}x_{\text{Ca}}^2$$
$$+ 1/2x_{\text{Fe}}x_{\text{Ca}} + x_{\text{Mg}}x_{\text{Fe}}x_{\text{Ca}}) + E_{\text{Fe-Ca}}(x_{\text{Fe}}^2x_{\text{Ca}} - x_{\text{Ca}}^2x_{\text{Fe}}) \quad \text{(VIII.22)}$$
$$- 1/2x_{\text{Fe}}x_{\text{Ca}}) + E_{\text{Ca-Fe}}(x_{\text{Fe}}x_{\text{Ca}}^2 - x_{\text{Fe}}^2x_{\text{Ca}} - 1/2x_{\text{Fe}}x_{\text{Ca}})$$
$$+ C(2x_{\text{Mg}}x_{\text{Fe}}x_{\text{Ca}} - x_{\text{Fe}}x_{\text{Ca}})$$

where the E's are for the binary solution constants and C a ternary constant as explained in Chapter II. The subscript M 2 has been dropped throughout. For $\ln f_{\text{Fe-M 2}}$ a similar expression can be written by replacing Mg by Fe, Fe by Ca and Ca by Mg in the equation above. Substituting for the activity coefficients in Eq. (VIII.20) an equation of the following form is obtained:

$$\ln K_{\text{D(VIII.18)}} = \ln K_{\text{VIII.18}} = \frac{A_{\text{0-M1}}}{RT}(X_1) - \frac{A_{\text{1-M1}}}{RT}(X_2) - E_{\text{Mg-Fe}}(X_3)$$
$$- E_{\text{Fe-Mg}}(X_4) \quad \text{(VIII.23)}$$
$$- E_{\text{Mg-Ca}}(X_5) - E_{\text{Ca-Mg}}(X_6) - E_{\text{Fe-Ca}}(X_7) - E_{\text{Ca-Fe}}(X_8) - C(X_9)$$

where X_1, X_2, etc. are obtained by appropriate substitutions, for example,

$$X_3 = (x_{\text{Fe}}^2 - 2x_{\text{Mg}}x_{\text{Fe}}^2 + 1/2x_{\text{Fe}}x_{\text{Ca}} - x_{\text{Mg}}x_{\text{Fe}}x_{\text{Ca}})$$
$$- (2x_{\text{Fe}}x_{\text{Mg}}^2 + 1/2x_{\text{Ca}}x_{\text{Mg}} + x_{\text{Mg}}x_{\text{Fe}}x_{\text{Ca}}) \quad \text{(VIII.24)}$$

where the terms in the first bracket are from Eq. (VIII.22) for $\ln f_{\text{Mg-M 2}}$ and those in the second from a similar equation for $\ln f_{\text{Fe-M 2}}$. This equation has ten constants and we need as many data on the composition of M 1 and M 2, preferably many more. The evaluation of the solution parameters would essentially depend on the number and accuracy of the data. Note that $(E_{\text{Mg-Fe}} + E_{\text{Fe-Mg}})/2$ is equal to $A_{\text{0-M 2}}/RT$ and $(E_{\text{Fe-Mg}} - E_{\text{Mg-Fe}})/2$ is equal to $A_{\text{1-M 2}}/RT$. In view of the large number of constants involved in Eq. (VIII.23) for the asymmetric model, it may be useful to try first simple mixture models for the two sites. This may be done by substituting $A_{\text{1-M 1}} = $ zero and $E_{\text{Mg-Fe}} = E_{\text{Fe-Mg}}$, $E_{\text{Mg-Ca}} = E_{\text{Ca-Mg}}$, $E_{\text{Fe-Ca}} = $ and $C = 0$ in Eq. (VIII.23). This reduces the number of constants to a manageable five.

The data needed for evaluating the solution parameters are the isothermal site occupancy of Mg, Fe and Ca in M 1 and M 2 sites in Ca-

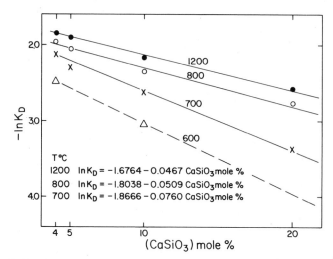

Fig. 46. The dependence of Mg–Fe distribution between M1 and M2 sites in calcic pyroxene on the concentration of Ca in M2 site. Data from unpublished results of SAXENA, TURNOCK and GHOSE. Symbols are for samples at different temperatures

pyroxenes which may vary in composition for all three components. Such data are not yet available.

The nature of mixing in M 2 site may be qualitatively inferred from the variation in the distribution coefficient $K_{D(VIII.18)}$ with changing concentration of Ca in M 2 site. HAFNER, VIRGO and WARBURTON's (1971) data on the Fe^{2+}–Mg order-disorder in the lunar clinopyroxenes show that increasing concentration of Ca in M 2 increases the preference of Fe^{2+} for that site. Some preliminary data from SAXENA, TURNOCK and GHOSE (unpublished) on the relationship between $\ln K_{D(VIII.18)}$ and the percent of $CaSiO_3$ in synthetic Ca-pyroxene is presented in Fig. 46. The site occupancy of Fe^{2+} and Mg in M 1 and M 2 sites in heated Ca-pyroxenes has been determined by Mössbauer spectroscopy. The details of the experiments are similar to those of SAXENA and GHOSE (1971). Fig. 46 shows that $-\ln K_D$ decreases linearly with increasing Ca in M 2 site indicating an increasing preference of Fe^{2+} over Mg for the site.

b) Thermodynamic Nature of Mixing in M 2 from Data on Orthoenstatite-Diopside Binodal

In Chapter VI, methods to calculate the thermodynamic functions of mixing from two-phase data are described. These methods are not applicable to the orthoenstatite-diopside solvus data since there is a structural change in going from orthoenstatite to diopside. These methods cannot be made applicable here by defining the components of the solution differently. One may consider defining components of a crystalline solution without necessarily having such components as discrete units. For example $MgSiO_3$ or MgO may be considered as components in pyroxenes. We have

$$\mu_{MgO\text{-}opx} = \mu^0_{MgO\text{-}En} + RT \ln x_{MgO\text{-}opx} . \qquad \text{(VIII.25)}$$

Note that $\mu^0_{MgO\text{-}En}$ is the chemical potential of MgO in pure enstatite. Free energy of enstatite is expressed as

$$G_{En} = \mu^0_{MgO\text{-}En} + \mu^0_{SiO_2\text{-}En} . \qquad \text{(VIII.26)}$$

In this equation the two chemical potentials may vary depending on the chemical composition and properties of coexisting phases (see RAMBERG, 1963). This is particularly true in the case of clinopyroxene where $\mu^0_{MgO\text{-}Di}$ is a function of P, T and composition. For clinopyroxene, the chemical potential of MgO is given by

$$\mu_{MgO\text{-}cpx} = \mu^0_{MgO\text{-}Di} + RT \ln a_{MgO\text{-}cpx} \qquad \text{(VIII.27)}$$

where $\mu^0_{MgO\text{-}Di}$ is given by

$$\mu^0_{MgO\text{-}Di} = G_{Di} - \mu^0_{CaO\text{-}Di} - 2\mu_{SiO_2\text{-}Di} . \qquad \text{(VIII.28)}$$

Note that

$$\mu^0_{MgO\text{-}En} \neq \mu^0_{MgO\text{-}Di} \qquad \text{(VIII.29)}$$

Similar arguments are valid when $MgSiO_3$ is used as a component instead of MgO.

It is evident that we are facing here a situation created by the unrealistic molecular treatment of an essentially ionic crystalline solution (ionic in this context does not imply that the bonds are purely ionic). As noted before, the atoms or ions may be treated as components and their chemical potential be defined by

$$\mu_i = \left(\frac{\partial G}{\partial n_i} \right)_{T, P, n_j}$$

where n represent the number of atoms or ions. Let us, therefore, consider that at equilibrium at a given P and T, we have

$$\mu_{Mg\text{-}opx} = \mu_{Mg\text{-}cpx} \qquad \text{(VIII.30)}$$

or

$$(\mu_{Mg\text{-}M\,1} + \mu_{Mg\text{-}M\,2})_{opx} = (\mu_{Mg\text{-}M\,1} + \mu_{Mg\text{-}M\,2})_{cpx} \qquad \text{(VIII.31)}$$

where $\mu_{\text{Mg-M 1}}$ and $\mu_{\text{Mg-M 2}}$ are chemical potentials on the non-equivalent sites and have been defined in Chapter VII. In the present context some more explanation of these potentials is necessary.

As discussed by MUELLER, GHOSE and SAXENA (1971) and in Chapter VII, in classical thermodynamics. two different chemical potentials for a species on two nonequivalent sites in a single homogeneous phase are inadmissable. One may, however, following BORGHESE (1967), consider in the orthoenstatite-diopside crystalline solution, Mg on M 1 site as a different species from Mg on M 2 site since in orthoenstatite M 1 site is geometrically different from M 2 site and both are completely occupied by Mg. Furthermore Mg on M 1 site in diopside should also be regarded as a different species from Mg on M 1 site in orthoenstatite because the energetic properties of the former are influenced by the presence of Ca in M 2 and, therefore, they must be different from the latter.

In BORGHESE's (1967) terms, one may talk about the chemical potential of the site M 2 referred to Mg in the crystalline solution, and about the chemical potentials of the site M 1 in orthoenstatite and of the site M 1 in diopside both referred to Mg.

In the orthoenstatite-diopside series, M 1 is completely occupied by Mg and, therefore, Mg and Ca mixing is confined to the site M 2. The geometric changes in the M 2 polyhedra due to the changing Mg/Ca ratio cause the changes in the geometry of the M 1 polyhedra and the latter may be regarded as a necessary consequence of the former. At equilibrium in the coexisting phases on the binodal solvus we have besides (VIII.30):

$$(\mu_{\text{Mg-M 2}})_{\text{opx}} = (\mu_{\text{Mg-M 2}})_{\text{cpx}} , \qquad \text{(VIII.32)}$$

$$(\mu_{\text{Mg-M 1}})_{\text{opx}} = (\mu_{\text{Mg-M 1}})_{\text{cpx}} \qquad \text{(VIII.33)}$$

where $\mu_{\text{Mg-M 2}}$ and $\mu_{\text{Mg-M 1}}$ are the chemical potentials of M 2 and M 1 sites referred to Mg and may be expressed as:

$$(\mu_{\text{Mg-M 2}})_{\text{opx}} = (\mu^0_{\text{Mg-M 2}})_{\text{En}} + RT \ln (a_{\text{Mg-M 2}})_{\text{opx}} , \qquad \text{(VIII.34)}$$

$$(\mu_{\text{Mg-M 2}})_{\text{cpx}} = (\mu^0_{\text{Mg-M 2}})_{\text{Di}} + RT \ln (a_{\text{Mg-M 2}})_{\text{cpx}} , \qquad \text{(VIII.35)}$$

$$(\mu_{\text{Mg-M 1}})_{\text{opx}} = (\mu^0_{\text{Mg-M 1}})_{\text{En}} + RT \ln (a_{\text{Mg-M 1}})_{\text{opx}} , \qquad \text{(VIII.36)}$$

$$(\mu_{\text{Mg-M 1}})_{\text{cpx}} = (\mu^0_{\text{Mg-M 1}})_{\text{Di}} + RT \ln (a_{\text{Mg-M 1}})_{\text{cpx}} . \qquad \text{(VIII.37)}$$

$(\mu^0_{\text{Mg-M 1}})_{\text{En}}$ may be defined as the partial internal energy of M 1 site referred to Mg^{2+} in pure enstatite (see BORGHESE, 1967; TOLMAN, 1959) and can be regarded as the energy per Mg^{2+} ion in M 1 site, when this site is completely occupied by Mg^{2+} at the same P and T as in the actual orthopyroxene crystalline solution. The other chemical potentials are defined similarly. Note that for the species Mg-M 2 we may consider the same standard state throughout the solution series because the character of the M 2 site is directly a function of its occupancy by Mg or Ca. Therefore, we have

$$(\mu^0_{Mg\text{-}M\,2})_{En} = (\mu^0_{Mg\text{-}M\,2})_{Di} \qquad (VIII.38)$$

However, we have at the M 1 site

$$(\mu^0_{Mg\text{-}M\,1})_{En} \neq (\mu^0_{Mg\text{-}M\,1})_{Di} \qquad (VIII.39)$$

because the energetic behaviour of the site should be regarded as a function of the changing Mg/Ca ratio in M 2 site. As M 1 site is completely occupied by Mg the "partial" activities $a_{Mg\text{-}M\,1}$'s are indeterminate.

Combining Eqs. (VIII.38) and (VIII.39), we have

$$(\mu^0_{Mg\text{-}M\,2} + \mu^0_{Mg\text{-}M\,1})_{En} \neq (\mu^0_{Mg\text{-}M\,2} + \mu^0_{Mg\text{-}M\,1})_{Di} \qquad (VIII.40)$$

or

$$\mu^0_{Mg\text{-}En} \neq \mu^0_{Mg\text{-}Di}$$

which is consistent with the thermochemical data.

Although the species Mg-M 2 and Mg-M 1 are poorly defined in as much as Mg in M 1 or M 2 cannot have a separate existence of its own, it may be mentioned that the partial internal energy $\mu^0_{Mg\text{-}M\,1}$ or $\mu^0_{Mg\text{-}M\,2}$ have a definite meaning. This energy of a cation in the structural site is due to contributions of the various parameters such as (a) electrostatic Madelung potential (b) Born repulsion energy (c) ligand field stabilization (d) anion polarization (e) covalent bonding (f) vibrational and (g) Van der Waals energy.

With the above background, the mixing of Ca and Mg in M 2 site and the calculation of "partial" thermodynamic functions of mixing may be considered. However it should be emphasized that these "partial" quantities such as activity of an ion at a structural site cannot be considered to represent the activity of the ion in the crystal as a whole. In orthopyroxene it was possible to determine the "partial" activities of Fe and Mg simultaneously at the two sites and then combine them to represent the activity of Fe or Mg in the crystal as a whole. In enstatite-diopside series, this cannot be done since the Mg content of M 1 is stoichiometrically fixed. The "partial" functions of mixing of Mg and Ca at the M 2 site may be useful in understanding the nature of mixing in the crystal as a whole particularly since, in the present work, we are considering the changes in M 1 polyhedra from enstatite to diopside as a function of Ca/Mg ratio in M 2 site.

From (VIII.32), (VIII.34), (VIII.35) and (VIII.38), we have

$$RT \ln (a_{Mg\text{-}M\,2})_{opx} = RT \ln (a_{Mg\text{-}M\,2})_{cpx} \qquad (VIII.41)$$

WARNER (1971) has redetermined the enstatite-diopside solvus as a function of temperature at 2, 5 and 10 Kbar. Using these data and Eq.(VIII.41) and following the method of calculation as described in Chapter VI (Eqs.(VI.8) to (VI.11)), the two parameters A_0 and A_1 for Mg-Ca mixing on M 2 site can be calculated. WARNER (1972, personal communication) has supplied the data on the parameters W_{G_1} and W_{G_2} as a function of temperature and pressure. These data in terms of A_0 and A_1 are

$$A_0/RT = 31.710 \pm 1.847 - (11.024 \pm 1.349)\, T/1000 - (0.042 \pm 0.034)\, P$$

$$\text{(VIII.42)}$$

$$A_1/RT = 12.988 \pm 1.275 - (9.932 \pm 0.931)\, T/1000 - (0.150 \pm 0.050)\, P$$

$$\text{(VIII.43)}$$

where T is in °K and P in Kbar. Component 1 is Mg and component 2 is Ca in M 2 site.

With added information on the partial activities from the study of intracrystalline distribution, it should eventually be possible to obtain a complete information on the thermodynamic nature of the calcic pyroxenes.

IX. Olivine Crystalline Solution

Olivines $(Fe, Mg, Ca)_2 Si_2 O_4$ are important constituents of many igneous and metamorphic rocks. The thermodynamic properties of olivines have been studied recently by several workers (KITAYAMA and KATSURA, 1968; NAFZIGER and MUAN, 1967; OLSEN and BUNCH, 1970; WARNER, 1971). BOWEN and SCHAIRER (1935) calculated the molal heats of fusion for forsterite $(Mg_2 SiO_4)$ and fayalite $(Fe_2 SiO_4)$. SAHAMA and TORGESSON (1949) determined the heat of solution in HF for the crystalline solution series. The thermodynamic nature of the (Mg–Fe) olivine crystalline solution appears to be closely ideal at high temperatures from the data of Bowen and Schairer and from the cation distribution data (see for example OLSEN and BUNCH, 1970; MEDARIS, 1969). However, NAFZIGER and MUAN (1967) and KITAYAMA and KATSURA (1968) note a definite deviation from ideality in olivines at 1200° C. The results of the various workers are reviewed and some new calculations on activity-composition relations are presented in this chapter.

1. BOWEN and SCHAIRER's Data

Their data on liquidus and solidus curves are shown in Fig. 47. Ion-exchange reaction at a given temperature between olivine crystalline solution (s) and the liquid solution (1) may be expressed as

$$\underset{\text{Fo-s}}{Mg_2 SiO_4} + \underset{\text{Fa-1}}{Fe_2 SiO_4} \rightleftharpoons \underset{\text{Fo-1}}{Mg_2 SiO_4} + \underset{\text{Fa-s}}{Fe_2 SiO_4} \qquad \text{(IX.1)}$$

At equilibrium, we have

$$\mu_{\text{Fo-s}} = \mu_{\text{Fo-1}} \qquad \text{(IX.2)}$$

or

$$\mu_{\text{Fo-s}}^0 + 2RT \ln a_{\text{Fo-s}} = \mu_{\text{Fo-1}}^0 + 2RT \ln a_{\text{Fo-1}} \qquad \text{(IX.3)}$$

or

$$2RT \ln \frac{a_{\text{Fo-1}}}{a_{\text{Fo-s}}} = (\mu_{\text{Fo-s}}^0 - \mu_{\text{Fo-1}}^0) = \Delta \mu_{\text{Fo, s-1}}^0 . \qquad \text{(IX.4)}$$

Similarly we have

$$2RT \ln \frac{a_{\text{Fa-1}}}{a_{\text{Fa-s}}} = \Delta \mu_{\text{Fa, s-1}}^0 . \qquad \text{(IX.5)}$$

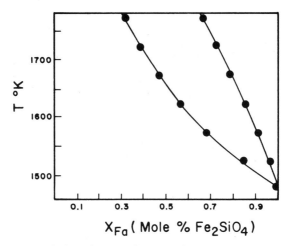

Fig. 47. Phase relations in Mg_2SiO_4–Fe_2SiO_4 sytem. Data from BOWEN and SCHAIRER (1935). Curves are fitted by assuming ideal solution in olivine (BRADLEY, 1962). The lower curve is the solidus; the upper the liquidus

The difference between the pure potentials of forsterite and fayalite in solid and liquid state may be expressed (see BRADLEY, 1962) as

$$\Delta \mu^0_{Fo, s-l} = (\Delta H_{Fo} - T_1 \Delta C_{P\text{-}Fo})(1/T_1 - 1/T) + \Delta C_{P\text{-}Fo} \ln T/T_1 \quad (IX.6)$$

where ΔH_{Fo} is the molar heat of fusion of forsterite at the melting point $T_1°K$ and $\Delta C_{P\text{-}Fo}$ is the increase in specific heat at constant pressure per mole on melting forsterite. Similarly $\Delta \mu^0_{Fa, s-l}$ may be expressed as

$$\Delta \mu^0_{Fa, s-l} = (\Delta H_{Fa} - T_2 \Delta C_{P\text{-}Fa})(1/T_2 - 1/T) + \Delta C_{P\text{-}Fa} \ln T/T_2 \quad (IX.7)$$

If olivine is assumed to be an ideal solution, the activities in Eqs. (IX.4) and (IX.5) are replaced by mole fraction as done by BOWEN and SCHAIRER (1935) and by BRADLEY (1962). BRADLEY's redetermination of heats of fusion for fayalite and forsterite are 25 200 and 29 300 cal/mole respectively. These differ from those of BOWEN and SCHAIRER's (~ 14000 cal/mole for both fayalite and forsterite) because BRADLEY rightly uses the ionic model and the specific heats in the calculation. It should, however, be mentioned that both in the works of BOWEN and SCHAIRER and BRADLEY, the latent heats of fusion are calculated by assuming that the solutions are ideal. Note that we cannot conclude from these data that the solutions are ideal as supposed by many authors including BRADLEY. As suggested by BOWEN and SCHAIRER the form of the liquidus and solidus curves appear to be similar to that formed by the ideal solutions.

2. Fe-Mg Olivines at 1200° C

The activity-composition relation at 1200° C in (Fe, Mg) olivine has been determined by NAFZIGER and MUAN (1967) and KITAYAMA and KATSURA (1968). As noted in Chapter V, NAFZIGER and MUAN used the reaction

$$(FeSi_{0.5}O_2) \text{ in olivine} = Fe + 1/2\, SiO_2 + 1/2\, O_2 \qquad (IX.8)$$

for which we have

$$\log a_{FeSi_{0.5}O_2} = \frac{\Delta G^0_{IX.8}}{2.303\, RT} + 1/2 \log P_{O_2} \qquad (IX.9)$$

where P_{O_2} is the oxygen pressure of the gas phase in equilibrium with the solid phases olivine crystalline solution, silica and metallic iron. If P_{O_2} is determined experimentally and $\Delta G^0_{(IX.8)}$, the change in the standard free energy for the reaction (IX.8), is known, the activity of fayalite in olivine crystalline solution can be estimated.

Another method to estimate the activity-composition relation is to first determine such a relation for pyroxene using a reaction similar to (IX.8) and then consider the reaction

$$(FeSiO_3)_{\text{in pyroxene}} + (MgSi_{0.5}O_2)_{\text{in olivine}}$$

$$\rightleftharpoons (MgSiO_3)_{\text{in pyroxene}} + (FeSi_{0.5}O_2)_{\text{in olivine}} \qquad (IX.10)$$

The equilibrium constant for (IX.10) is

$$K_{IX.10} = \frac{a_{MgSiO_3\text{-px}}\, a_{FeSi_{0.5}O_2\text{-ol}}}{a_{FeSiO_3\text{-px}}\, a_{MgSi_{0.5}O_2\text{-ol}}}. \qquad (IX.11)$$

If the activity-composition data are available in pyroxene, the activity coefficients for olivines may be determined by using the following equation (discussed before in Chapters I and V).

$$\ln f_{FeSi_{0.5}O_2} = -x_{MgSi_{0.5}O_2} \ln K_{10} + \int_0^{x_{MgSi_{0.5}O_2}} \ln K'_{IX.10}\, x_{MgSi_{0.5}O_2} \qquad (IX.12)$$

where K'_{10} is

$$\frac{a_{MgSiO_3\text{-px}}\, a_{FeSi_{0.5}O_2\text{-ol}}}{a_{FeSiO_3\text{-px}}\, a_{MgSi_{0.5}O_2\text{-ol}}}$$

Similarly for the other olivine component. NAFZIGER and MUAN (1967) found that olivine crystalline solution is somewhat non-ideal at 1200° C. (NAFZIGER and MUAN, 1967; LARIMER, 1968). The distribution data do not indicate a systematic variation of the distribution coefficient $K_D((x_{Mg\text{-px}} \cdot x_{Fe\text{-ol}})/(x_{Fe\text{-px}} \cdot x_{Mg\text{-ol}}))$ with the compositions of olivine and pyroxene. If the calculation of activities is attempted from these data by

Table 11. Distribution of Fe and Mg in coexisting olivine and pyroxene at 1200° C

S. No.	$x_{Mg\text{-}px}$	$x_{Mg\text{-}ol}$	$K_{D(IX.10)}$	Data
1	0.95	0.95	1.0	NAFZIGER and MUAN
2	0.89	0.89	1.0	(1967)
3	0.86	0.85	1.08	
4	0.78	0.74	1.24	
5	0.76	0.70	1.35	
6	0.71	0.64	1.37	
7	0.60	0.57	1.13	
8	0.60	0.55	1.22	
9	0.55	0.51	1.17	
10	0.48	0.44	1.17	
11	0.44	0.38	1.28	
12	0.983	0.982	1.06	LARIMER (1968)
13	0.955	0.950	1.12	
14	0.944	0.940	1.07	
15	0.927	0.917	1.15	
16	0.856	0.837	1.16	
17	0.694	0.652	1.21	

$K_{D(IX.10)}$ is the ideal distribution coefficient. Pyroxene is closely ideal (W/RT or $\alpha = 0.07 \pm 0.12$, NAFZIGER and MUAN). Olivine may also be regarded as closely ideal on the basis of the present data at 1200° C.

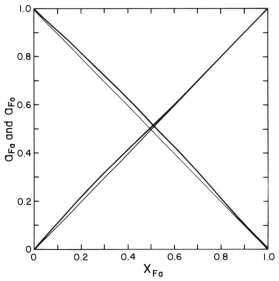

Fig. 48. Activity-composition relation in olivine at 1200° C as calculated from LARIMER's (1968) data, assuming orthopyroxene as ideal. The deviations from ideality are small and olivine may well be regarded as ideal at 1200° C

using one of the Eqs.(IV.11), (V.6) or (V.14), the results appear to be unrealistic. As for example, the calculations from NAFZIGER and MUAN's (1967) data using Eq.(V.6) show that for pyroxene and olivine A_0's are 1.74 and 1.89 and A_1's -0.65 and -0.54 respectively. As compared to other results, these values predict too much non-ideality of solution in olivine. Table 11 shows that the variation of $K_{D(IX.10)}$ with x_{Mg} either in olivine or pyroxene is systematic only for the first six samples in NAFZIGER and MUAN's data. If LARIMER's (1968) data are used for calculating activities using the two parameter Eq.(V.6) with the assumption that pyroxene is ideal, parameters obtained for olivine solution are $A_{0/RT} = 0.124$ and $A_{1/RT} = -0.072$ (cal/mole) while $\ln K = 0.254$. The parameters are for components 1 as $MgSi_{0.5}O_2$ and 2 as $FeSi_{0.5}O_2$ and, therefore, a negative A_1 implies that higher excess free energy of mixing is associated with the fayalite rich side. These activity-composition relations are shown in Fig. 48 and show that the deviation from ideality in olivine (Mg$_2$SiO$_4$–Fe$_2$SiO$_4$) at 1200° C is small.

3. The System Monticellite (CaMgSiO$_4$)– Forsterite (Mg$_2$SiO$_4$)

WARNER (1972, personal communication) determined the binary solvus data on the CaMgSiO$_4$–Mg$_2$SiO$_4$ join and calculated the thermodynamic functions of mixing in the crystalline solution using the one parameter (symmetric) Margules equation. WARNER has found that the constant W_G is a function of P and T and has presented least squares equations relating W_G to P and T. WARNER's results are

$$5\,\text{Kbar} \quad W_G = \underset{\pm 4402}{35000} - \underset{\pm 6752}{37247(T\,1000^{-1})} + \underset{\pm 2557}{13094(T\,1000^{-1})^2}\,, \quad \text{(IX.13)}$$

$$10\,\text{Kbar} \quad W_G = \underset{\pm 6931}{46391} - \underset{\pm 10672}{55466(T\,1000^{-1})} + \underset{\pm 4050}{20396(T\,1000^{-1})^2}\,. \quad \text{(IX.14)}$$

Temperature is $T(°K)$ and W_G is cal/mole. The estimated standard errors in W_G are 195.2 at 5 Kbar and 341.7 at 10 Kbar.

A further least squares analysis of the polybaric polythermal data gives the following general equation

$$W_G = \underset{\pm 2531}{24786} - \underset{\pm 3672}{19605\,(T\,1000^{-1})} + \underset{\pm 1324}{5588\,(T\,1000^{-1})^2}\,, \qquad \text{(IX.15)}$$

$$\underset{\pm 93.3}{- 380.9\,P} + \underset{\pm 69.7}{327.8\,P\,(T\,1000^{-1})}\,.$$

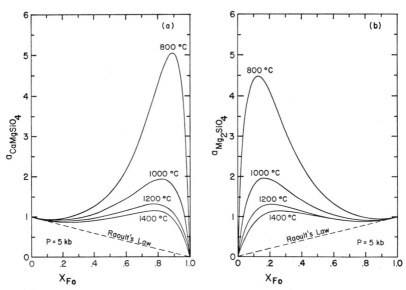

Fig. 49. Activity-composition relation in olivine ($CaMgSiO_4$–Mg_2SiO_4). (After WARNER, 1971)

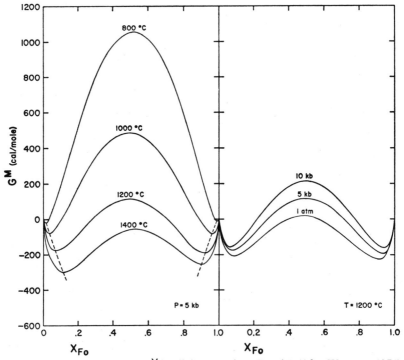

Fig. 50. Free energy of mixing G^M in olivine at various T and P. (After WARNER, 1971)

P is in kilobars and T and W_G are as in (IX.13) and (IX.14). The estimated standard error in W_G is 316.9 cal/mole.

For these Eqs. (IX.13) to (IX.15) component 1 is $CaMgSi_2O_4$ and component 2 is Mg_2SiO_4. These equations are sufficient to calculate A_0 and therefore any thermodynamic function of mixing at any P and T. As examples, the activity-composition relations between 800 to 1400° C at 5 Kbar have been shown in Fig. 49 and the free energy of mixing G^M at different temperatures and pressures in Fig. 50.

X. Feldspar Crystalline Solution

Although important information on crystal structure and experimental phase relationship has been accumulating for decades, it is only recently that attempts have been made to use such information in understanding the thermodynamic properties of the pure end member feldspar or their crystalline solutions. ALLMANN and HELLNER (1962), PERCHUK (1965), PERCHUK and RYABCHIKOV (1968), THOMPSON and WALDBAUM (1968a, 1968b, 1969), and THOMPSON (1969) have discussed and calculated the energy functions of feldspar solutions. WALDBAUM (1968) has calculated the thermodynamic properties of alkali feldspars while HOLM and KLEPPA (1968) and BROWN (1971) have calculated the ideal configurational entropies for plagioclase. Most of the energy calculations for the feldspar crystalline solution have been based on the experimental data on phase relationships as collected by ORVILLE (1963), BOWEN and TUTTLE (1950) and LUTH and TUTTLE (1966). Recently ORVILLE (1972), BACHINSKI and MULLER (1971) and SECK (1971a) have published more experimental data on binary and ternary feldspars.

In this chapter, we shall be concerned with the three feldspars, albite $NaAlSi_3O_8$, anorthite $CaAl_2Si_2O_8$, and K-feldspar $KAlSi_3O_8$. The symbols Ab, An, and Or would represent the three molecules $NaAlSi_3O_8$, $CaAl_2Si_2O_8$ and $KAlSi_3O_8$.

1. Order-Disorder

Order-disorder in feldspar differs from order-disorder in pyroxenes in some important respects. The Si–O framework remains intact in pyroxenes while $Mg–Fe^{2+}$ order-disorder takes place on the non-equivalent metal sites. In feldspars the ions K^+ and Na^+ occupy equivalent sites but Si–Al order-disorder takes place in the Si–Al–O framework sites. While Mg^{2+} and Fe^{2+} may completely replace each other in pyroxenes the Si–Al replacement in alkali feldspar is stoichiometrically restricted to one Al and three Si atoms. In plagioclase the Si : Al ratio changes from 3 : 1 in albite to 2 : 2 in anorthite. Here, however, unlike that in pyroxene, the Al–Si order-disorder at a given temperature is a function of the concentration of albite and anorthite in the plagioclase crystalline solution.

a) Potassium Feldspars

There are two tetrahedral groups in sanidine T_1 and T_2, which are identical in size. Each site, therefore, may be assumed to contain (on the average) $Al_{1/4}Si_{3/4}$, the Al–Si distribution being completely random. This distribution varies continuously as a function of temperature. Partially ordered monoclinic potassium feldspars may be referred to as orthoclase. Table 12 shows the distribution of Al and Si between T_1 and T_2 in several different K-feldspars. The data are from PHILLIPS and RIBBE (1973).

Table 12. Distribution of Al between T_1 and T_2 in five K-Feldspars. Data from PHILLIPS and RIBBE (1973)

Specimen	Heated C	7002	C	7007	B
$Al(T_1)$	0.260	0.291	0.332	0.396	0.385
$Al(T_2)$	0.229	0.211	0.153	0.107	0.112
Total Al	0.489	0.502	0.485	0.503	0.497

Specimen C and B are Spencer C and Spencer B respectively. Specimen 7002 and 7007 are Laacher-See low sanidine and adularia (see HOVIS, 1971).

Microcline is triclinic and T_1 sites are replaced by pairs of $T_1(O) + T_1(m)$ sites and T_2 sites by pairs of $T_2(O) + T_2(m)$ sites. In intermediate microcline a typical distribution of Al may be $T_1(O)$ 0.65 $T_1(m)$ 0.25, $T_2(O)$ 0.03 and $T_2(m)$ 0.01 while in maximum microcline it may be $T_1(O)$ 0.93, $T_1(m)$ 0.03 $T_2(O)$ 0.01 $T_2(m)$ 0.01 (see TAYLOR, 1962).

b) Sodium Feldspar

The distribution of Al and Si among the sites $T_1(O)$, $T_1(m)$, and $T_2(O)$ and $T_2(m)$ in triclinic albite changes continuously as a function of temperature from low albite to high albite. The following distribution of Al atoms in albite is given by STEWART and RIBBE (1969).

	Site				
	$T_1(O)$	$T_1(m)$	$T_2(O)$	$T_2(m)$	Total
Low albite	0.88	0.02	0.06	0.05	1.02
High albite	0.28	0.25	0.22	0.25	1.00

c) Calcium Feldspar

Primitive anorthite structure is the typical structure for low-temperature anorthite. It, however, seems to persist even at higher temperature (RIBBE and MEGAW, 1962). The cell contains 16 crystallo-graphically distinct tetrahedra, eight small and eight large. The mean T–O distances correspond to nearly complete ordering.

d) Process of Ordering

Even with all the progress in the crystal-structural and chemical work on the feldspars, the process of ordering is incompletely understood in detail. Broadly speaking the distribution of Al and Si is a continuous function of temperature. Sodium feldspar at room temperature is triclinic but it is likely that a truly monoclinic phase exists at temperatures near the melting point. As we cool the crystal, Al tends to become concentrated in the $T_1(O)$ site by a regular migration of Al from T_2 sites and $T_1(m)$ into $T_1(O)$ site. Fig. 51 (STEWART and RIBBE, 1969) shows the Al distribution over the four tetrahedral sites in albite as a function of temperature. As ordering proceeds on cooling from the disordered state, the Al concentration in $T_1(O)$ and $T_1(m)$ first diverge linearly at a low rate. Below about 550° C the rate of divergence of the Al concentration in the two sites increases markedly. Above 550° C the Al contents

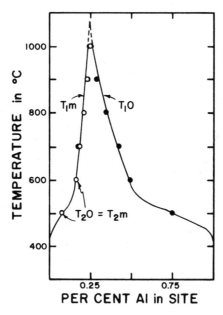

Fig. 51. Al distribution over the four tetrahedral sites in albite as a function of temperature, as deduced by STEWART and RIBBE (1969)

of $T_1(m)$, $T_2(O)$ and $T_2(m)$ are approximately equal. Below $550°$ C the Al content of $T_1(m)$ may be slightly less than that of the T_2 sites. With gentle heating we shall have the migration of Al from T_1O site to the other three sites. If this addition of Al to the three sites is more or less equal above $550°$ C, it is evident that in order for albite to become monoclinic we must have complete disorder, i.e. equal site occupany in all the four sites which may take place at or near the liquidus temperature.

If similar ordering process operates in K-feldspars, we may expect a "truly" monoclinic symmetry only when there is complete disorder in sanidine. It does seem as remarked by STEWART and RIBBE (1969), that there are triclinic domains in sanidine that result from the random migration of Al into the equivalent $T_1(O)$ and $T_1(m)$ sites. Although K-feldspar begins to order with monoclinic restraints, but on unit-cell scale these restraints rapidly disappear as twin-related triclinic domains nucleate and grow, eventually giving way to topologically distinct macroscopic units. Thus it would appear that the process of ordering in both Na- and K-feldspar is similar. However, the kinetics of ordering is different since in K-feldspar the early formed twin-related domains must be recrystallized to macroscopic domains. Under identical physical conditions, the rate of ordering in K-feldspar is, therefore, slower than that in albite where the macroscopic domains form directly without the need to recrystallize microscopic domains.

2. Long-Range Ordering Parameters and the Calculation of Al Site-occupancy from Crystal-structural Data

A triclinic alkali feldspar has four distinct tetrahedral sites and it requires three site occupancies to describe the order-disorder completely. THOMPSON (1969) suggested the use of an ordering parameter Z defined as

$$Z = x_{(T_1,O)Al} + x_{(T_1,m)Al} - x_{(T_2,O)Al} - x_{(T_2,m)Al} \qquad (X.1)$$

where x_{Al} is the site occupancy of Al in the tetrahedral site specified. For monoclinic feldspars, because of the structural equivalence of T_1O to T_1m and T_2O to T_2m, is given by:

$$Z = 2(x_{(T_1)Al} - x_{(T_2)Al}). \qquad (X.2)$$

Z equal to zero corresponds to complete disorder and Z equal to unity to complete order. This brings us to the problem of determining the site occupancy of Al in the tetrahedral sites.

By using the data in Table 12, PHILLIPS and RIBBE (1973) suggest the use of the following equations in estimating the Al–Si distribution in monoclinic feldspars:

$$Al(T_1) = 2.360\,(c - 0.4b) - 4.369 \tag{X.3}$$
$$Al(T_2) = 2.256\,(c - 0.4b) + 4.658 \tag{X.4}$$

where b and c represent the dimensions of the b and c cell edges. Earlier STEWART and RIBBE (1969) found a linear correlation between Δbc, which is a proportional distance, and Al site occupancy. They suggest the estimation of Al site occupancy in alkali feldspar of any structural state by estimating Al in T_1 from Δbc, obtaining Al in T_2 by difference, and distributing Al between T_1 sites from $\Delta(\alpha^*\gamma^*)$, where

$$Al_{T_1(O)} = \frac{\Delta(bc) + \Delta(\alpha^*\gamma^*)}{2} \quad \text{and} \quad Al_{T_1(m)} = Al_{T_1(O)} - \Delta(\alpha^*\gamma^*) \tag{X.5}$$

The Al in T_2 sites can be divided equally between $T_2(O)$ and $T_2(m)$.

The quantities $(c - 0.4b)$ and Δbc are often referred to as structural ordering parameters. Unfortunately the values of the Al site occupancy used in arriving at the relationships between Al site occupancy and the parameters Δbc and $(c - 0.4b)$ are themselves determined by assuming a linear relationship between the site occupancy and the mean T–O bond length (SMITH and BAILEY, 1963; RIBBE and GIBBS, 1969). Therefore, it should be emphasized that there is no way as yet to determine definite quantitative Al–Si distribution and any calculations of thermodynamic quantities based on such data are only tentative.

Using PHILLIPS and RIBBE's results, we may use the following relationship to obtain the ordering parameter Z from b and c cell edges:

$$Z = -18.054 + 9.232\,(c - 0.4b)$$

3. Ideal Configurational Entropy

The distribution of Al on any one of the two sites T_1 and T_2 can be regarded as random when the site occupancy of Al in the site approaches zero or unity. There may or may not be a random distribution in other cases when the site occupancy of Al is significantly different from zero or unity. The ideal configurational entropy for a monoclinic feldspar may be calculated using the following equation:

$$S^{IM} = -2R\left[x_{AlT_1}\ln x_{AlT_1} + x_{SiT_1}\ln x_{SiT_1} + x_{AlT_2}\ln x_{AlT_2} + x_{SiT_2}\ln x_{SiT_2}\right] \tag{X.6}$$

Entropies of variously disordered K-feldspars and Na-feldspars have been calculated by HOLM and KLEPPA (1968), BROWN (1971) and HOVIS (1971). As mentioned before, the site occupancy determination using the methods discussed in the previous section, are not accurate enough for a meaningful thermodynamic analysis. Examples of such entropy values and the effect of the uncertainties of Al site occupancies on these values have been given by HOVIS (1971).

One may calculate the ideal configurational entropy of mixing for the plagioclase crystalline solutions which may be close to the true entropy of mixing at high temperatures ($\sim 800°$ C). In plagioclase, however, the mixing involves the exchange of CaAl and NaSi and we must take into account the contributions to the entropy of mixing both from the mixing of Na–Ca and Si–Al. For the configurational entropy in plagioclase we have

$$S^{IM} = -R\{x \ln x + (1-x)\ln(1-x)\} - 4$$
$$\cdot R \left\{ \frac{(2-x)}{4} \ln \frac{(2-x)}{4} + \frac{(2+x)}{4} \ln \frac{(2+x)}{4} \right\} \qquad (X.7)$$

where x is mole fraction $\dfrac{Na}{Na+Ca}$. It may be noted that $(1-x)$ is $\dfrac{Na}{Na+Ca}$ and the quantities $\dfrac{(2-x)}{4}$ and $\dfrac{(2+x)}{4}$ would correspond to site occupancies of Al and Si respectively. The terms in the first brackets are for the mixing of Na and Ca and in the second bracket for the mixing of Al and Si. It is important that we define the mole fraction x as $Na/(Na+Ca)$ and not as $Ab/(Ab+An)$. The latter would mean that we are considering the mixing of the molecules Ab ($NaAlSi_3O_8$) and An ($CaAl_2Si_2O_8$). The entropy of mixing in the molecular model would be different from that of the ionic model. In the former case there is no separate entropy contribution due to Al–Si mixing since such mixing necessarily follows that of Ab and An.

For a disordered plagioclase with equal numbers of Na and Ca ions ($x = 0.5$) we have

$$S^{IM} = -R(1/2 \ln 1/2 + 1/2 \ln 1/2) - 4R(3/8 \ln 3/8 + 5/8 \ln 5/8)$$
$$= 6.64 \text{ cal/deg/mole}$$

which is similar to the calculations of BROWN (1971) except that here the mole fraction x is defined in terms of ions and not in terms of Ab and An. The ideal configurational entropies for the disordered pure end members Ab and An are given by

$$S^{IM}(Ab) = -4R(1/4 \ln 1/4 + 3/4 \ln 3/4) = 4.47 \text{ cal/deg/mole},$$
$$S^{IM}(An) = -4R(1/2 \ln 1/2 + 1/2 \ln 1/2) = 5.52 \text{ cal/deg/mole}.$$

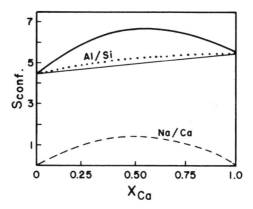

Fig. 52. Ideal configurational entropy in plagioclase. The solid line denotes total entropy while the dotted and dashed lines indicate the contribution of changing Al/Si and Na/Ca respectively. Fig. modified from BROWN (1971)

Fig. 52 shows the variation in the total entropy in the disordered plagioclase solution. The contribution of Na–Ca and Al–Si mixing to the total entropy is also shown. The figure is from BROWN (1971) but unlike that in BROWN, x is Na/(Na + Ca).

4. Thermodynamic Functions of Mixing in Binary Feldspars

a) Monoclinic Alkali Feldspar

A major contribution towards understanding the thermodynamic properties of alkali feldspar is that of THOMPSON and WALDBAUM (1968a, 1968b, 1969a, 1969b) and of WALDBAUM and THOMPSON (1969). Their calculations of the thermodynamic functions of mixing are based on the experimental data of ORVILLE (1963) and LUTH and TUTTLE (1966). The theory and method of such calculations have already been presented in Chapter VI. In order that these experimental data be suitable for such calculations, we must be certain that chemical equilibrium was closely approached between the coexisting phases. THOMPSON and WALDBAUM chose their data carefully. However, there are two problems which require discussion. The first problem is that of polymorphism and the second of Al–Si order-disorder. These problems arise because of the reason discussed below.

Consider two coexisting feldspars on the binodal solvus at a given temperature. If the two feldspars with components Ab and Or have approached equilibrium, we have:

$$\mu_{Ab-\alpha} = \mu_{Ab-\beta}, \tag{X.8}$$

$$\mu_{Or-\alpha} = \mu_{Or-\beta} \tag{X.9}$$

where α is the albite rich phase and β the K-feldspar rich phase, and

$$\mu_{Ab-\alpha} = \mu_{Ab}^0 + RT \ln a_{Ab-\alpha}, \tag{X.10}$$

$$\mu_{Ab-\beta} = \mu_{Ab}^0 + RT \ln a_{Ab-\beta} \tag{X.11}$$

where μ_{Ab}^0 is the chemical potential of pure albite. Let us now suppose that pure albite and pure K-feldspar have different structures at the given temperature T. In such a case the structures of α and β are also likely to be different. It is evident that it will be erroneous to consider a triclinic pure albite to be a component of monoclinic K-feldspar rich phase since such a triclinic component cannot really be present in the structure. It is evident that if we like to eliminate the constants such as μ_{Ab}^0 from (X.8) we must have that α and β obey the same equation of state. This has been clearly stated and discussed by THOMPSON and WALDBAUM (1969a). Similar considerations show that not only the structures of both albite and K-feldspar should be monoclinic but at the given temperature, the Al–Si ordering should also be identical.

Unfortunately at the present state of our ability to measure the Al–Si ordering in feldspars, we must assume that the monoclinic albite and K-feldspar do not differ in their Al–Si ordering more significantly than the errors involved in the measurement of Al site occupancy. At high temperatures (above $\sim 620°$ C at 2 Kbar) the crystalline solutions and the pure end members are monoclinic in the experiments. THOMPSON and WALDBAUM (1969b) give the following equation for the molar excess Gibbs energy:

$$G^{EM}(P, T) = (6326.7 + 0.0925\,P - 4.6321\,T)\,x_{Ab}\,x_{An}^2$$
$$+ (7671.8 + 0.1121\,P - 3.8565\,T)\,x_{An}\,x_{Ab}^2 \tag{X.12}$$

where T is temperature (°K) and P denotes pressure (bars). The numerical coefficients (Margules parameters) are based on data ranging from 500 to 700° C and from 2–10 Kbar. The calculations of thermodynamic functions of mixing using such parameters should be regarded as "qualitatively like the correct ones and quantitatively not far off the mark".

To be consistent throughout this book, instead of the formulation with Margules parameter, we shall continue with the parameters A_0 and A_1 which are given by

$$A_0 = 6999.25 + 0.1023\,P - 4.2443\,T, \tag{X.13}$$

$$A_1 = 672.55 + 0.0098\,P - 0.3878\,T. \tag{X.14}$$

The excess heat and entropy of mixing are given by (II.25) and (II.24) respectively. Volume of mixing may also be calculated by using an expression similar to (II.24) and differentiating A_0 and A_1 for P instead of T.

Figs. 53 and 54 (after WALDBAUM and THOMPSON, 1969) show the free energy of mixing and activity of Ab and An components. Spinodal curve passes through points of inflection and all compositions between binodal and spinodal curves are metastable with respect to diffusion. The critical constants of unmixing for disordered monoclinic feldspars

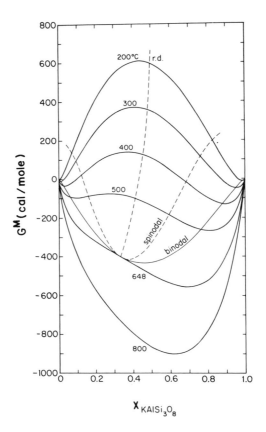

Fig. 53. Isobaric-polythermal projection of Gibbs energy of mixing surface of sanidine solutions at 1 bar. Trace of binodal curve on the surface is shown by a light solid line. Spinodal curve (dashed line) passes through points of inflection on the isothermal sections of the surface. All compositions between binodal and spinodal curves are metastable with respect to diffusion. Reference state is a mechanical mixture of pure end-member phases at the stated temperature and 1 bar.
Fig. after WALDBAUM and THOMPSON (1969)

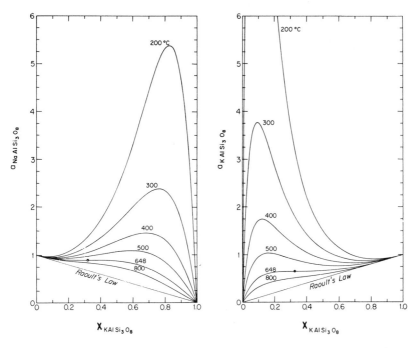

Fig. 54. Activities of (a) $NaAlSi_3O_8$ and (b) $KAlSi_3O_8$ in sanidine crystalline solutions at 1 bar. Reference states are the pure endmember phases at the stated temperature and 1 bar. Solid circles denote critical point. Fig. after WALDBAUM and THOMPSON (1969 b)

are given by

$$T_c(°K) = 921.23 + 13.4607\, P_c(Kbar),\qquad(X.15)$$

$$x_{or} = 1/3.\qquad(X.16)$$

These results apply only to highly disordered monoclinic feldspars. The binodal solvus becomes different for ordered triclinic feldspars (BACHINSKI and MÜLLER, 1971). The critical constants of unmixing for such feldspars at atmospheric pressure are:

$$T_c = 885 \sim 887° C,$$
$$X_{or} = 0.24 \sim 0.27.$$

Ordered feldspars with compositions on such a binodal solvus at high temperatures are metastable. These results, therefore, are of only theoretical interest.

b) Plagioclase

Recently SECK (1971a) has presented data on coexisting ternary feldspars in the system Ab–An–Or between 650 and 900° C. In addition the compositions of coexisting plagioclase and Na–Ca chloride solution at 700° C and 2 Kbar have been published by ORVILLE (1972). The results that follow are based on these data.

Consider the following reaction between coexisting sanidine (s) and plagioclase (pl), assuming that sanidine is a binary solution of the end members molecule albite (Ab) and orthoclase (Or) and that plagioclase is a binary solution of albite and anorthite (An):

$$x_{\text{Ab-s}} = x_{\text{Ab-pl}} \tag{a}$$

where x_{AB} is $\text{Ab}/(\text{Ab} + \text{Or})$. Note that unlike other ion exchange relationships, e.g. as that between Mg^{2+} and Fe^{2+} silicates, reaction (a) involves a change in the relative amounts of the phases present.

At a given P and T,

$$\mu_{\text{Ab-s}} = \mu_{\text{Ab-pl}} \tag{X.17}$$

or

$$\mu_{\text{Ab}}^0 + RT \ln a_{\text{Ab-s}} = \mu_{\text{Ab}}^0 + RT \ln a_{\text{Ab-pl}} \tag{X.18}$$

where μ_{Ab}^0 is the chemical potential of pure albite. We then have

$$RT \ln \frac{a_{\text{Ab-pl}}}{a_{\text{Ab-s}}} = RT \ln \frac{x_{\text{Ab-pl}} f_{\text{Ab-pl}}}{x_{\text{Ab-s}} f_{\text{Ab-s}}} = 0, \tag{X.19}$$

which may be rewritten in the form (see Chapter V):

$$RT \ln \frac{x_{\text{Ab-pl}}}{x_{\text{Ab-s}}} = A_{0\text{-s}} x_{\text{Or-s}}^2 + A_{1\text{-s}} (3 x_{\text{Ab-s}} - x_{\text{Or-s}}) x_{\text{Or-s}}^2 - A_{0\text{-pl}} x_{\text{An-pl}}^2 \tag{X.20}$$
$$- A_{1\text{-pl}} (3 x_{\text{Ab-pl}} - x_{\text{An-pl}}) x_{\text{An-pl}}^2 .$$

It should be emphasized that both alkali feldspar and plagioclase are assumed to be binary solutions. If, however, the concentration of An in alkali feldspar and concentration of Or in plagioclase are significant, the equilibrium conditions are

$$\mu_{\text{Ab-s}} = \mu_{\text{Ab-pl}}$$
$$\mu_{\text{Or-s}} = \mu_{\text{Or-pl}} \tag{X.21}$$
$$\mu_{\text{An-s}} = \mu_{\text{An-pl}} .$$

For ternary solutions the series expansion for the activity coefficients or the excess free energy of mixing in terms of mole fractions may contain as many as seven or nine constants (see Chapter V). SECK's (1971a) data contain compositions of coexisting ternary alkali feldspar and plagioclase.

However, at 650° C there is little solution of anorthite in alkali feldspar (1.5–2.0%) and it may be neglected. In plagioclase, particularly at the albite end, there is a significant concentration of orthoclase (2.0 to 9.5%). Thus a ternary solution model is more appropriate for plagioclase. There are data only on 13 pairs of coexisting feldspars with which to solve an equation with 9 unknowns. A good solution is, therefore, not possible and we may only calculate the energy constants pertaining to the binary joins albite-anorthite and albite-orthoclase.

SECK's data for 650° C are presented in Table 13. The mole fractions x_{Ab-pl} and x_{Ab-s} are recalculated assuming each phase to be a binary solution. Since the concentration of Or is significant in sodic plagioclase, the mole fractions x_{Pl-Ab} are subject to systematic error. The effect of this error on the calculation of activity cannot be evaluated without using a ternary model. However, it is possible to compare our results on the Ab–Or join with those of THOMPSON and WALDBAUM (1969a). This would give us the idea of the errors that may be involved in the calculations for plagioclase feldspar.

Using Eq. (X.20), we arrange the data in a matrix form and use a standard computer program for matrix solution. Table 14 shows the constants A_{0-s}, A_{0-pl}, A_{1-s} and A_{1-pl} for the two feldspars at 650° C. The standard statistical errors on these constants are negligible. Note also that the constants A_{0-s} and A_{1-s} compare well with such constants determined by THOMPSON and WALDBAUM (1969a).

Table 13. Compositions of coexisting feldspars at 1 Kbar and 650° C. Data from SECK (1971a)

Alkali feldspar			Plagioclase		
x_{Ab}	x_{Or}	x_{An}	x_{Ab}	x_{Or}	x_{An}
0.475	0.511	0.015	0.838	0.095	0.067
0.435	0.546	0.020	0.819	0.076	0.105
0.445	0.536	0.020	0.833	0.081	0.086
0.374	0.607	0.020	0.800	0.072	0.129
0.323	0.658	0.020	0.761	0.067	0.172
0.343	0.637	0.020	0.790	0.067	0.143
0.297	0.688	0.015	0.742	0.062	0.196
0.256	0.729	0.015	0.703	0.058	0.240
0.215	0.770	0.015	0.624	0.053	0.323
0.189	0.796	0.015	0.560	0.039	0.402
0.168	0.817	0.015	0.490	0.034	0.476
0.153	0.833	0.015	0.440	0.029	0.531
0.126	0.864	0.015	0.374	0.020	0.607

We may use yet another method and data for calculating A_0 and A_1 for plagioclase crystalline solution. ORVILLE (1972) has published data on the compositions of the equilibrium products in the reaction:

$$2NaAlSi_3O_8 + CaCl_2 = CaAl_2Si_2O_8 + 2NaCl + 4SiO_2 \tag{b}$$

Here we must consider the mixing of Ab and An in the crystalline solution and of [$CaCl_2$] and [$2NaCl$] in the vapor phase. The vapor phase of the chloride solution has been shown to be ideal by ORVILLE (1972); therefore, the equilibrium constant for reaction (b) is:

$$K_b = \frac{a_{An} \, x_{NaCl}^2 a_{SiO_2}^4}{x_{Ab}^2 \, x_{CaCl_2}} \frac{f_{An}}{f_{Ab}^2} \tag{X.22}$$

where x_{Ab} and x_{An} are mole fractions Ab/(Ab + An) and An/(Ab + An), as before, and X_{NaCl} and X_{CaCl_2} are the atomic ratios Na/(Na + Cl) and Ca/(Ca + Na) respectively in the vapor phase.

In the equilibrium products, quartz is always present. Setting a_{SiO_2} as unity and taking the logarithm of both sides, Eq. (X.22) can be written as:

$$\ln K_b = \ln K_D + \ln \left(\frac{f_{An}}{f_{Ab}} \right) - \ln f_{Ab} \tag{X.23}$$

where

$$K_D = \frac{x_{An} \cdot x_{NaCl}^2}{x_{Ab}^2 \, x_{CaCl_2}} \tag{X.24}$$

Table 14. Solution parameters A_0 and A_1 for alkali and plagioclase feldspars

	Ab-Or cal/mol.	Ab-An cal/mol.	
A_0	3920	1320	SECK's data at 650° C and 1 Kbar;
A_1	657	373	binary model
A_0	3508		Ion-exchange data at 650° C and 1 Kbar;
A_1	675		THOMPSON and WALDBAUM (1969)
A_0		967	ORVILLE's (1972) ion-exchange data
A_1		715	at 700° C and 2 Kbar

Note: The true intrinsic error in the calculations of A_0 and A_1 are unknown. Statistical errors in the constants from SECK's data are insignificant. For the constants from ORVILLE's data we have A_0 as 967 ± 87 and A_1 as 715 ± 100.

Using relations (V.3) and (V.4) the activity coefficients may be expressed in terms of mole fractions, producing the following equation:

$$\ln K_b = \ln K_D + \frac{A_0}{RT}(x_{Ab} - x_{An} - x_{An}^2)$$

$$+ \frac{A_1}{RT}(6\,x_{Ab}\,x_{An} - 1 + 3\,x_{An}^2\,x_{Ab} - x_{An}^3) \tag{X.25}$$

A multivariate linear regression analysis (Biomedical computer programs, multiple regression, Dixon, 1970) on the data collected in the form as required by Eq. (X.25), yields an intercept, which is the value of $\ln K_b$ and two regression coefficients which are A_0/RT and A_1/RT (Table 14). As in the previous case, the standard errors of regression coefficients (see Dixon, 1970, p. 258) do not cover any systematic errors. Fig. 55 represents the values of calculated $\ln K_D$ plotted against x_{Ab}.

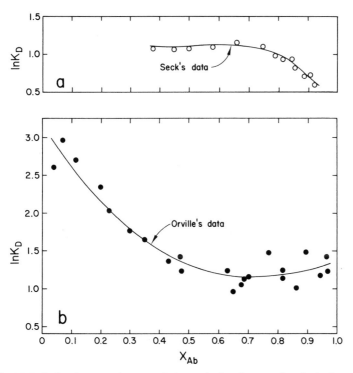

Fig. 55. Relation between $\ln K_D$ and the mole fraction x_{Ab} in plagioclase. K_D for Seck's data is $x_{Ab\text{-}pl}/x_{Ab\text{-}s}$ and K_D for Orville's data is $x_{An}x_{NaCl}^2/x_{Ab}^2 x_{CaCl_2}$. The curves are obtained by least squares analysis according to Eq. (X.20) for Seck's data and according to Eq. (X.25) for Orville's data

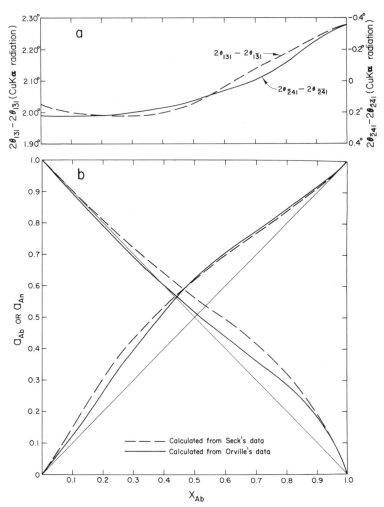

Fig. 56. Activity-composition relation in plagioclase crystalline solution. The dashed curve is calculated using Seck's data (650° C, 1 kbar). The solid curve is calculated using Orville's data (700° C, 2 Kbar). Lattice parameters in Fig. (a) for high plagioclase. Note the continuous variation in these parameters with composition

The values for the constants A_0 and A_1 for plagioclase obtained from Seck's data differ from those calculated by using Orville's data. Although pressure and temperature were somewhat different in the two cases (650° C at 1 Kbar and 700° C at 2 Kbar), Seck's (1971b) results indicate that the compositions at 700° C and 2 Kbar are nearly equivalent to those at 685° C and 1 Kbar. Fig. 56 shows the activity-composition

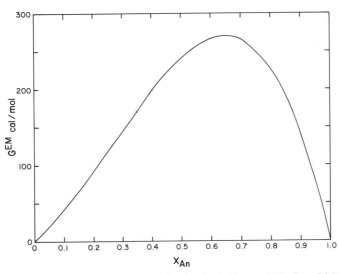

Fig. 57. Excess free energy of mixing in plagioclase at 700° C and 2 Kbar

relations in plagioclase crystalline solution calculated by using the two sets of data listed in Table 14. The differences between the two sets of values for A_0 and A_1 may be due to the fact that SECK's feldspars are ternary solutions.

The activity-composition relations in both cases show that the crystalline solution is somewhat asymmetric. In other words, as shown in Fig. 57 the higher values of excess free energy of mixing are associated with anorthite rich feldspars. A structural explanation of the energy variation is proposed by SAXENA and RIBBE (1973). The lattice parameter of high plagioclase indicate that from An_0 with its disordered 7 Å, $C\bar{1}$ cell, completely random substitution of Al for Si occurs in the four tetrahedral sites at least up to the An_{50} composition (see RIBBE, STEWART, and PHILLIPS, 1970). Evidently in this range the excess free energy of mixing is mainly due to the mixing of Na and Ca. From An_{100}, with its ordered 14 Å, $I\bar{1}$ (or $P\underline{\bar{1}}$) structure, substitution of Si for Al is quite likely governed by the Al-avoidance principle (LOEWENSTEIN, 1954; LAVES and GOLDSMITH, 1955) and occurs at the four (or eight) Al sites, perhaps down to the An_{85} composition. In the range An_{50} to An_{85} the substitution is more complex because of the transition between a 7Å and a 14Å framework. Any high plagioclase between these compositions would show long range disorder of Al–Si when averaged onto the four T sites of 7Å, $C\bar{1}$ cell; however, this region may well be characterized by an increasing degree of short range order which is fully developed only at An_{100}, the composition at which short- and long-range order become equivalent (see SMITH, 1970; RIBBE, 1972). In this composition range, therefore, we may expect that the excess free energy of mixing would be higher than in any other range as noted in Fig. 57.

c) Ternary Feldspars

At 900° C there is significant ternary solution in feldspars as seen in SECK's (1971a) data. These data, however, could not be used to obtain ternary solution parameters by regression analytical methods of Chapter V. Some semi-quantitative information on the parameters of binary solution anorthite-orthoclase and, therefore, on the nature of the ternary solution may be obtained by fitting a binodal curve by trial and error to the data on composition of coexisting feldspars at 900° C. Let it be assumed that the binary solution parameters A_0 and A_1 (and therefore E_{ij}'s) do not change significantly with temperature for the binary systems albite-anorthite ($E_{23} = 300$, $E_{32} = 1700$ cal/mol) and albite-orthoclase ($E_{13} = 4183$, $E_{31} = 2833$ cal/mol) as determined before for

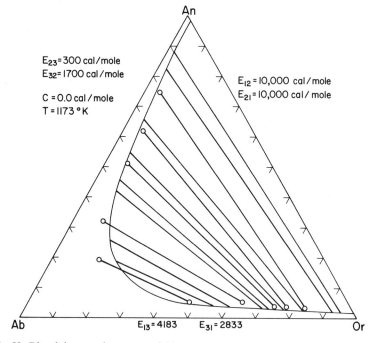

Fig. 58. Binodal curve in ternary feldspar. The open circles are composition of coexisting feldspars at 900° C (SECK, 1971). The solid line is the theoretical curve fitted with the constants shown in the figure by trial and error

the data at $650°$ C (see Table 14, note that $A_0 = (E_{ij} + E_{ji})/2$ and $A_1 = (E_{ji} - E_{ij})/2$). Using the same constants at $650°$ C for the data at $900°$ C and fixing the constants for the binary system anorthite-orthoclase by trial and error, the binodal fit to the data at $900°$ C is obtained as shown in Fig. 58. Note that the fit is particularly bad towards the high albite feldspars. The manipulation of the constants for the binary anorthite-orthoclase alone does not yield a better fit than that shown in the figure. It also appears from SECK's data that the binary solution anorthite-orthoclase is slightly asymmetric. It may, however, be suggested tentatively that this binary solution approximates closely to a regular solution ($A_0 = W = 10000$ cal/mol).

XI. Crystalline Solutions and Geothermometry

The concept of metamorphic facies evolved through the attempts of the petrologists to distinguish the mineral assemblages formed at different P and T in the field. In several cases, the physical and chemical conditions of the formation of rocks have been simulated in the laboratory. However, the meagerness of relevant thermodynamic data on rock-forming minerals makes it difficult to interpret meaningfully and check the validity of most experiments. From existing knowledge, it is possible to obtain only certain qualitative to semiquantitative estimates of P and T of the formation of a mineral assemblage. Such methods are based on the knowledge of the chemical reactions that occur as a result of changing P and T within and in between the crystalline solutions.

This section is not a review of the methods of geothermometry. INGERSON (1955) and SMITH (1963) have made an excellent review of such methods. Coexisting crystalline solutions and their thermodynamic properties of mixing as a function of P and T shall be the main concern of this chapter.

1. Intercrystalline Equilibria

a) Distribution of a Component between Coexisting Simple Mixtures

Assume that data on the distribution of a component A between two binary phases (A, B) M and (A, B) N exist. The distribution data cover a large compositional range in both the phases. Assume that the activity-composition relations of (A, B) M, a simple mixture are known at several different temperatures but that the activity-composition relation in (A, B) N is unknown. If the form of the distribution isotherm, whose temperature is unknown, does not indicate (A, B) N to be too nonideal, the distribution can be assumed to have the form:

$$\ln K = \ln K_D + \frac{W_\alpha}{RT}(1 - 2x_{Fe-\alpha}) - \frac{W_\beta}{RT}(1 - 2x_{Fe-\beta}), \qquad (XI.1)$$

where K and K_D are the equilibrium constant and the distribution coefficient, respectively, and α and β represent (A, B) M and (A, B) N, respectively.

To find the temperature of the distribution isotherm (considering that the effect of P is negligible on the equilibrium constant $K_{XI.1}$)

the temperature may be assumed to be T_1 and Eq. (XI.1) may be solved by a least-squares method. Temperatures T_2, T_3, etc., would also be assumed and the procedure repeated. In practice, geologic evidence would fix the limits T_1 and T_2 within which the actual temperature of the ion exchange equilibrium might lie. These calculations yield a set of W_α and W_β values at temperatures T_1 and T_2. Such values of W_α at T_1 and T_2 differ from the experimental values of W_α at these temperatures because the temperature of the isotherm is different from the assumed T_1 or T_2. The calculated values of W_α and the previously calculated values of W_α from experimental results may be plotted against T, as in Fig. 59. The intersection of the two lines yields the temperature of the distribution isotherm.

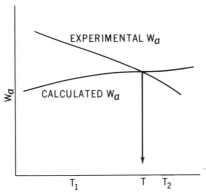

Fig. 59. Estimation of temperature of crystallization. See text for explanation

Unfortunately most minerals are not binary and many of those that are quasi-binary may not be strictly symmetric simple mixtures. The method, however, may be found useful in some cases. Consider for example the distribution of Fe^{2+} and Mg between M1 sites in orthopyroxene and clinopyroxene, assuming that in the latter mineral M2 site is completely occupied by Ca. The distribution data in some metamorphic rocks of granulite facies was presented in Table 1 [1]. Corresponding to each mole fraction of Fe in orthopyroxene X_{Fe-opx}, data on M1 site occupancy can be obtained from SAXENA and GHOSE (1971) (see Chapter VIII) at different temperatures. Such compositions were also listed in Table 1 and are plotted against the mole fraction X_{Fe-cpx} in clinopyroxene in Fig. 60. The open circles in Fig. 60 are distribution points based on the assumption that the temperature of intercrystalline ion exchange equilibrium between M1 site in orthopyroxene and M1 site in clinopyroxene was 600° C. Similarly the solid circles represent

---

[1] see SAXENA (1971)

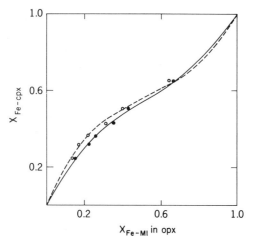

Fig. 60. The distribution Fe and Mg between M1 in opx and cpx. The curves are least squares fit according to Eq. (XI.3)

700° C. The M1 site in orthopyroxene is a simple mixture of Fe and Mg. Assuming that the one site quasi-binary clinopyroxene (M2 occupied by Ca) is also a simple mixture, the log of the equilibrium constant for the ion exchange

$$\underset{\text{opx}}{Mg(M1)} + \underset{\text{cpx}}{Fe(M1)} \rightleftharpoons \underset{\text{opx}}{Fe(M1)} + \underset{\text{cpx}}{Mg(M1)} \qquad (XI.2)$$

is given by

$$\ln K_{XI.2} = \ln K_{D(XI.2)} + \frac{W_{M1}}{RT}(1 - 2x_{Fe\text{-}M1}) - \frac{W_{cpx}}{RT}(1 - 2x_{Fe\text{-}cpx}) \qquad (XI.3)$$

where K_D is as before $(x_{Fe\text{-}M1} \, x_{Mg\text{-}cpx})/(x_{Mg\text{-}M1} \, x_{Fe\text{-}cpx})$. Fig. 60 shows distribution curves drawn by using a least squares fit to the distribution data according to Eq. (XI.3). The resulting constants are

	$K_{XI.2}$	W_{cpx}/RT	W_{M1}/RT
600° C	0.81	0.35	1.01
700° C	0.88	0.58	1.07

From SAXENA and GHOSE (1971), the correct values of W_{M1}/RT in orthopyroxenes are 1.38 and 0.99 Kcal/mole at 600 and 700° C respectively. The difference between these values and the calculated values of W_{M1}/RT here is due to the actual temperature of crystallization being different from either 600 or 700° C. This temperature is 680° C as determined graphically (see Fig. 59) from the intersection of the straight lines, one joining the calculated W_{M1}/RT and the other joining the experimental W_{M1}/RT. The success of this method depends on the accuracy of the compositional data because the W's are quite sensitive to any significant changes in the mole fractions. In the present case the estimated temperature for the metamorphic pyroxenes is quite reasonable.

b) Coexisting Plagioclase and Alkali Feldspar

BARTH (1956, 1962) proposed that the ratio of the mole fraction of albite in alkali feldspar to the mole fraction of albite in plagioclase varies linearly with the inverse of absolute temperature. This is true if mixing of the end members Ab and orthoclase $[Or (KAlSi_3O_8)]$ and Ab and anorthite $[An (CaAl_2Si_2O_8)]$ is ideal or nearly ideal in the compositional range.

The properties of sanidine (s) crystalline solutions have been studied by THOMPSON and WALDBAUM (1969b) and by THOMPSON (1969). Even in the high temperature range from 973–1273 °K, the solution deviates significantly from the ideal mixture. Similarly it is expected that there will be significant deviation from an ideal state in the plagioclase feldspars.

Plagioclase and sanidine may be considered as two ternary solutions at a certain P and T, with ion-exchange reactions

$$\overset{pl}{NaAlSi_3O_8} \rightleftharpoons \overset{s}{NaAlSi_3O_8}, \tag{XI.4}$$

$$KAlSi_3O_8 \rightleftharpoons KAlSi_3O_8, \tag{XI.5}$$

and

$$CaAl_2Si_2O_8 \rightleftharpoons CaAl_2Si_2O_8. \tag{XI.6}$$

The corresponding equilibrium constants are

$$K_{XI.4} = \frac{a_{Ab\text{-}s}}{a_{Ab\text{-}pl}}$$

$$= \frac{x_{Ab\text{-}s} f_{Ab\text{-}s}}{x_{Ab\text{-}pl} f_{Ab\text{-}pl}} \tag{XI.7}$$

$$K_{XI.5} = \frac{a_{Or\text{-}s}}{a_{Or\text{-}pl}}$$

$$= \frac{x_{Or\text{-}s} f_{Or\text{-}s}}{x_{Or\text{-}pl} f_{Or\text{-}pl}} \tag{XI.8}$$

and

$$K_{XI.6} = \frac{a_{An\text{-}s}}{a_{An\text{-}pl}}$$

$$= \frac{x_{An\text{-}s}\, f_{An\text{-}s}}{x_{An\text{-}pl}\, f_{An\text{-}pl}} \tag{XI.9}$$

where $x = Ab/(Ab + An + Or)$. The activity coefficients f are functions of P, T, and the ratio of two of the other mole fractions.

In actual practice, the situation may be somewhat simplified by assuming that both sanidine and plagioclase are binary solutions. Therefore,

$$K_{XI.4} = \frac{x_{Ab\text{-}s}\, f_{Ab\text{-}s}}{x_{Ab\text{-}pl}\, f_{Ab\text{-}pl}}$$

where $x_{Ab\text{-}s} = Ab/(Ab + Or)$ in sanidine and $x_{Ab\text{-}pl} = Ab/(Ab + An)$. $f_{Ab\text{-}s}$ is now a function of P, T, and x's, and $f_{Ab\text{-}pl}$ is similarly defined. Even in the simplified case, the activity-composition relations in both the sanidine and plagioclase solutions must be known. Activity-composition data are available for sanidine crystalline solutions (THOMPSON and WALDBAUM, 1969b) but for plagioclase crystalline solutions, the data are incomplete. Only sanidine and high-temperature plagioclase have been considered above. The temperature estimate at lower ranges of temperatures is further complicated by the structural changes in the feldspars as a function of T and composition.

SECK's data (1971a) on ternary coexisting feldspars between 500 bars to 1 Kbar and 650 to 900° C may be used to construct a feldspar geothermometer by simply representing SECK's data in a more convenient diagram as discussed below. Consider an ion exchange between coexisting feldspars as

$$Or\text{-}s + Ab\text{-}pl \rightleftharpoons Ab\text{-}s + Or\text{-}pl \tag{XI.10}$$

for which the distribution coefficient is

$$K_{D(XI.10)} = \frac{x_{Ab\text{-}s}\, x_{Or\text{-}pl}}{x_{Ab\text{-}pl}\, x_{Or\text{-}s}} \tag{XI.11}$$

This distribution coefficient varies with temperature, pressure and the three mole fractions. From SECK's data, the functional dependence of K_D on temperature and composition is known. Therefore, SECK's data may be represented as shown in Fig. 61 where $\ln K_{D(XI.10)}$ is plotted against $x_{An\text{-}pl}$. One may also choose x_{Ab} or x_{Or} in either plagioclase or sanidine to be plotted on the abscissa. The isotherms are given by the

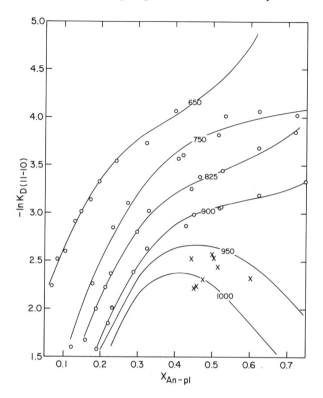

Fig. 61. A geothermometer based on SECK's (1971a) data on composition of coexisting feldspar at 650—900° C. The curves are least squares fitted according to the Eqs. (XI.12) to (XI.14). Crosses are compositions of Ischia lava. (RAHMAN and MACKENZIE, 1969)

following equations:

$$-\ln K_D \text{ at } 650° C = -1.40393 - 15.2168\, x_{\text{An-pl}}$$
$$+ 33.3632\, x^2_{\text{An-pl}} - 29.0077\, x^3_{\text{An-pl}},$$

$$-\ln K_D \text{ at } 750° C = .201974 - 18.8000\, x_{\text{An-pl}}$$
$$+ 29.3212\, x^2_{\text{An-pl}} - 15.8911\, x^3_{\text{An-pl}},$$

$$-\ln K_D \text{ at } 825° C = .721593 - 20.6703\, x_{\text{An-pl}}$$
$$+ 36.0606\, x^2_{\text{An-pl}} - 22.5520\, x^3_{\text{An-pl}},$$

$$-\ln K_D \text{ at } 900° C = 1.39942 - 21.7623\, x_{\text{An-pl}}$$
$$+ 35.7960\, x^2_{\text{An-pl}} - 20.2737\, x^3_{\text{An-pl}}.$$

(XI.12)

No correction for the difference in pressure (500 bars at 900° C and 1 Kbar at other temperatures) has been made. The first two constants in Eq. (XI.12) vary systematically with temperature and the relation is given by

$$
\left.
\begin{array}{l}
\text{First constant} = -22.794596 + 0.048880\,T - 0.0000245\,T^2 \\
\text{Second constant} = 39.937800 - 0.1271953\,T - 10.0000651\,T^2
\end{array}
\right\} \quad \text{(XI.13)}
$$

Unfortunately the third and fourth constants do not vary systematically with temperature. Linear relationships between temperature and the two latter constants are given by least squares analysis as follows

$$
\left.
\begin{array}{l}
\text{Third constant} = 21.4761 + 0.0155618\,T \\
\text{Fourth constant} = -42.7006 + 0.026585\,T
\end{array}
\right\} \quad \text{(XI.14)}
$$

It is possible to extrapolate and obtain isotherms at other temperatures as shown in Fig. 61. This figure may be used to find the temperature of the crystallization of the coexisting feldspars ($P \simeq 0.5$–1 Kbar). It should be possible to construct similar temperature-composition diagrams at other pressures by using SECK's data (1971b) on the effect of pressure on the ternary feldspar solvus. Fig. 61 shows the estimate of temperature for coexisting feldspars from trachytes in Ischia (RAHMAN and MACKENZIE, 1969). These estimated temperatures 950 to 1000° C are probably somewhat high.

c) Distribution of Fe and Mg in Coexisting Minerals

The use of the distribution coefficient involving two multicomponent minerals in problems of petrogenesis is severely limited. The distribution of Fe^{2+} and Mg^{2+} between coexisting garnet and biotite (bi) is one example. Such a distribution is a function of P, T, and the concentrations of Mn and Ca in garnet and of Al^{3+}, Fe^{3+}, and Ti in biotite (KRETZ, 1959; ALBEE, 1965; SEN and CHAKRABORTY, 1968; SAXENA 1968c).

Fortunately, the garnet-biotite pair has been the subject of much research, and several chemical analyses are available in the literature. It is, therefore, possible to improve the usability of K_D by analyzing its compositional dependence by a multivariate statistical analysis. SAXENA (1969b) used principal component analysis and obtained the transformed distribution coefficient as

$$
\begin{aligned}
\text{Transformed } K_D = {} & 0.5013\,K_D - 0.4420\,x_{\text{Fe-gar}} \\
& + 0.1506\,x_{\text{Fe-bi}} - 0.3474\,x_{\text{Mn-gar}} \\
& + 0.0865\,x_{\text{Ca-gar}} - 0.0333\,x_{\text{Al(IV)-bi}} - 0.3165\,x_{\text{Al(VI)-bi}} \\
& + 0.5488\,x_{\text{Ti-bi}}
\end{aligned} \quad \text{(XI.15)}
$$

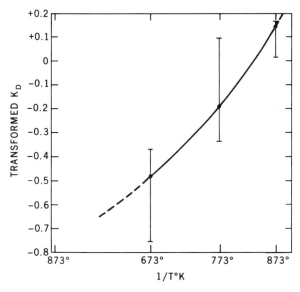

Fig. 62. Estimation of approximate temperature of crystallization using the Mg–Fe^{2+} distribution between garnet and biotite. The bars represent the spread in the "transformed K_D" values from rocks of the same metamorphic grade

where x_{Fe} is $Fe/(Fe + Mg)$, K_D is $x_{Fe\text{-}bi} x_{Mg\text{-}gar}/x_{Fe\text{-}gar} x_{Mg\text{-}bi}$, and the other x's represent ions on a 12–0 basis in garnet and 22–0 basis in biotite. The transformed K_D could be shown to vary systematically with temperature inferred on geologic evidence. Fig. 62 shows the use of such a transformed K_D in estimating the temperature of crystallization of rocks containing garnet and biotite. Using this method of estimating temperature, DENNEN, BLACKBURN and QUESADA (1970) obtained an average temperature of 923° K for certain Grenville gneisses, which agrees well with their independent estimate based on the concentration of Al^{3+} in quartz. Some of the garnet compositions, particularly from low-grade metamorphic rocks, may represent zoning. Therefore, there is further possibility of improving the quality of the transformed K_D as a temperature indicator. It is assumed that a P variation between 4 to 8 Kbars does not change K_D and other substitutional relations significantly.

Distribution of Fe and Mg between coexisting orthopyroxene and clinopyroxene has been investigated by several workers since KRETZ (1961, 1963) demonstrated that the distribution in coexisting pairs from igneous rocks is distinctly different from that in such pairs from the metamorphic rocks. Fig. 63 shows the distribution data as given by

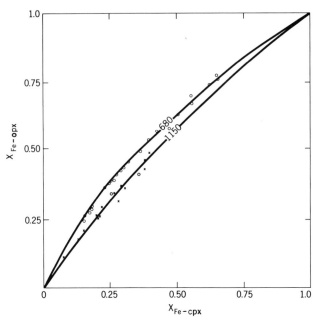

Fig. 63. Distribution of Fe and Mg between ortho- and calcic-pyroxene. The data are from KRETZ (1963). Open circles — metamorphic, crosses — igneous. The 680° C isotherm is based on activity-composition relations in clinopyroxenes as mentioned in the text and in orthopyroxene calculated by the analytical expression given in Chapter VIII. K is 0.61. The 1150° C isotherm is based on ideal solution model with K as 0.73

KRETZ (1963) and two isotherms corresponding to metamorphic (680° C) and igneous ($\approx 1150°$ C) rocks. The isotherm for the data from metamorphic rocks has been drawn by considering Ca-pyroxene as simple mixture ($W_{cpx}/RT = 0.54$, SAXENA, 1971) and using the activity-composition relation in orthopyroxene from SAXENA and GHOSE (1971). The isotherm for igneous pyroxenes is assumed to be ideal. The separation between the two isotherms, although distinct, is rather narrow which indicates that the temperature dependence of the equilibrium constant for Fe–Mg exchange is not large. Furthermore as the temperature increases, the concentration of Ca in both ortho- and clino-pyroxene changes which would affect the energetic properties of the crystals. Therefore no quantitative use of such distribution data can be made until the thermodynamic properties of the ternary pyroxene solution have been determined.

Distribution of Fe and Mg between coexisting garnet and Ca-pyroxenes has been studied by BANNO (1970) and SAXENA (1968b, 1969b).

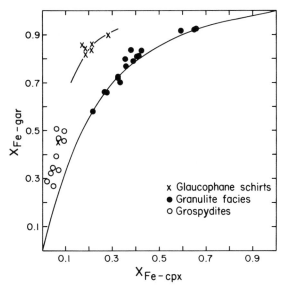

Fig. 64. Distribution of Fe–Mg between garnet and clinopyroxene. The curves are for ideal solutions in both Mg–Fe garnet and clinopyroxene

Some distribution data (SAXENA, 1969b) on the ion exchange

$$1/3Fe_3Al_2Si_3O_{12} + MgCaSi_2O_6 \rightleftharpoons 1/3Mg_3Al_2Si_3O_{12} + FeCaSi_2O_6 \quad (XI.16)$$
$$\underset{\text{Fe-gar}}{} \quad \underset{\text{Mg-cpx}}{} \quad \underset{\text{Mg-gar}}{} \quad \underset{\text{Fe-cpx}}{}$$

are presented in Fig. 64. The problems in using the distribution coefficient $K_{D(XI.16)}$ $(=(x_{Mg\text{-}gar})(x_{Fe\text{-}cpx})/(x_{Fe\text{-}gar})(x_{Mg\text{-}cpx}))$ as a P and T indicator are several. First, garnets contain at least four major components Fe, Mg, Ca and Mn and variation of Ca and Mn affects the energetic properties of the solution which probably also change somewhat as a function of Fe/Mg ratio. Second, clinopyroxene in many of the metamorphic rocks such as those from certain eclogites and from glaucophane schist facies contains a high solution of jadeite molecule. Third, in absence of the volume of mixing data on both garnet and clinopyroxene, the pressure dependence of the distribution coefficient cannot be accurately estimated. With all these uncertainties it is evident that a quantitative use of the distribution coefficient is not possible. However, the use of $K_{D(XI.16)}$ may be made in suitable cases to indicate the approximate P and T conditions. BANNO (1970), for example, uses $K_{D(XI.16)}$ to distinguish between various eclogites and finds that $K_{D(XI.16)}$'s for low temperature eclogites (California) amphibolite facies eclogites (Norway) and eclogites from kimberlites (Robert Victor Mine) are

0.044, 0.11 and 0.28 respectively. The following are empirically determined factors on the interrelationship between K_D, P, T and composition (BANNO, 1970; SAXENA, 1969b):

(1) $K_{D(XI.16)}$ decreases with increasing pressure.

(2) $K_{D(XI.16)}$ increases with increasing temperature.

(3) Compositional influence of Mn and Ca in garnet on $K_{D(XI.16)}$ is significant. The effect of changing Fe/Mg in both the minerals is unknown but is likely to be small.

The effect of changing jadeite component in pyroxene on $K_{D(XI.16)}$ is also unknown. However it must be noted that although these compositional influences on $K_{D(XI.16)}$ are significant in any quantitative estimates of P and T, they are not significant enough to prohibit a qualitative use of $K_{D(XI.16)}$ in distinguishing rocks formed under as different physical conditions as those of glaucophane schists and granulite facies.

d) Distribution of Other Elements

The experimental results of BUDDINGTON and LINDSLEY (1964) on the composition of coexisting iron-titanium oxides may be used to calculate the temperatures and oxygen fugacities of oxide equilibration as done by CARMICHAEL (1967).

It is expected that the distribution of some minor or trace components between coexisting phases may be particularly useful as P and T indicators. The concentration of such components are subject to large errors of determination. Accurate data are needed on the distribution of such elements as Ni, V, Co and Ba between coexisting minerals in rocks or in experimental assemblages. An interesting use of the distribution data on Fe, Mg and Co between coexisting olivine and orthopyroxene has been made by MATSUMOTO (1971) in estimating P and T of South African kimberlite and Ichinomegata Iherzolite.

Distribution of fluorine between coexisting phases has been considered as a potential P and T indicator by STORMER and CARMICHAEL (1971) and EKSTRÖM (1972).

2. Order-Disorder

a) Fe–Mg Silicates

The distribution of a cation over nonequivalent structural sites is a function of temperature and composition and in suitable cases, such distribution coefficients may be useful to understand the thermal history of the host rocks. The use of the intracrystalline distribution data, however, requires some additional considerations not required in the use of the intercrystalline data. These considerations involve the rate of the ion exchange process which is rapid as compared to the intercrystalline ion exchange. The order-disorder process may also have some potential barriers which may require very high activation energy to cross. VIRGO and HAFNER (1969) made the important observation that there is an apparent cutoff or transition region on the temperature scale below which no more ordering or disordering occurs. This transition temperature in orthopyroxene was estimated to be approximately 750° K. Above this temperature the activation energy required for diffusion to start in the direction of disordering is of the order of 20 kilocalories (VIRGO and HAFNER, 1969). Below this temperature the activation energy should be very high. This is confirmed from the measurement of order-disorder in metamorphic pyroxenes that cooled slowly through geological time. Fig. 65 shows the data on the K_D values for the distribution of Fe^{2+} and Mg^{2+} between M1 and M2 sites in metamorphic

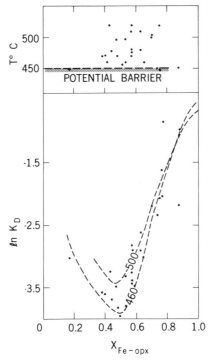

Fig. 65. Temperature estimate of the ion-exchange equilibrium between sites in natural orthopyroxenes. 450° C appears to be the transition temperature below which no more ordering takes place due to a potential barrier

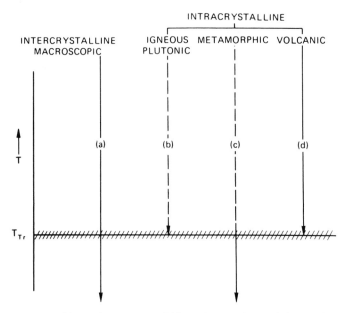

Fig. 66. Scheme illustrating the possibility of crystals attaining and retaining equilibrium distributions. The full lines indicate temperature ranges for different materials in which such distributions are readily attained and retained. Dashed lines indicate temperature ranges in which these possibilities are greatly diminished. Fig. after MUELLER (1970)

orthopyroxene. From these data it may be noted that no orthopyroxene shows a degree of order representing temperatures lower than 723° K.

Fig. 66 (from MUELLER, 1970) shows schematically the possibility of crystals attaining and retaining equilibrium cation distribution at different temperatures. The intercrystalline ion exchange equilibrium between two crystals may be attained and quenched in at any temperature at which crystallization or recrystallization occurred. Such a temperature represents the latest of the igneous or metamorphic event in which the new crystals were formed. An intercrystalline equilibrium between two pyroxenes established in plutonic norites at solidus temperatures does not change with the slow cooling and remains fixed unless there may be a complete recrystallization as under the conditions of granulite facies when the norites may be metamorphosed into charnockites. In Fig. 66 a solid line (a) indicates that the intercrystalline equilibrium may be attained and retained at all temperatures. On the other hand, the intracrystalline exchange is a rapid process and, therefore, the intracrystalline equilibrium in a phase crystallized at solidus temperature in igneous plutonic rock, continuously changes its site occupancies with the slowly falling temperature until the temperature falls below the cut-off or transition temperature (T_{Tr}, line b). Similarly in metamorphic rocks (line c) the ion exchange would continue to be adjusted with cooling until the transition temperature. If the rocks form below the temperature T_{Tr} by recrystallization, the intracrystalline equilibrium would persist unchanged with cooling because of the high activation energy required for the ordering process below T_{Tr}. In volcanic rocks because of rapid cooling the intracrystalline equilibrium may be quenched in somewhat below the actual temperature of crystallization of the phases.

b) Fe–Mg Order-Disorder in Orthopyroxene

The isotherms in Fig. 35 (Chapter VIII) may be used to estimate the temperature of intracrystalline ion exchange equilibrium in orthopyroxene. Fig. 67 shows a geothermometer which is based on the same

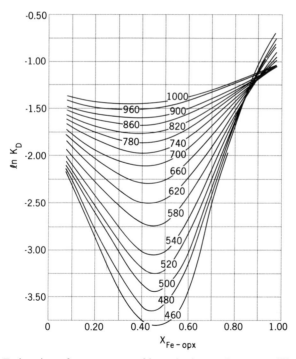

Fig. 67. Estimation of temperature of intersite ion-exchange equilibrium in orthopyroxene. K_D is $x_{Fe-M1}\, x_{Mg-M2}/x_{Mg-M1}\, x_{Fe-M2}$. In certain rocks quenched quickly such as some meteorites and volcanics, the temperature estimated by this method is close to the temperature of rock formation

data as in Fig. 35 with extrapolated values for other temperatures. From the discussion in the preceding section it is obvious that the temperature estimated from this figure is the temperature at which the ion exchange reaction ceased. For metamorphic pyroxenes this temperature will be in the vicinity of the transition temperature but for volcanic rocks or for some quenched meteorites, the estimated temperature would be close to the temperature of crystallization.

c) Order-Disorder in Feldspars

The ion exchange of Al and Si between T1 and T2 sites in a monoclinic feldspar may be expressed as

$$Al(T2) + Si(T1) \rightleftharpoons Al(T1) + Si(T2) \qquad (XI.17)$$

The distribution coefficient for the above reaction is

$$K_{D(XI.17)} = \frac{x_{Al\text{-}T1}\, x_{Si\text{-}T2}}{x_{Al\text{-}T2}\, x_{Si\text{-}T1}}. \qquad (XI.18)$$

Following THOMPSON (1969), a long-range ordering parameter was defined in the last chapter as

$$Z = 2(x_{Al\text{-}T1} - x_{Al\text{-}T2})$$

Since this parameter can be calculated directly from lattice parameters, we may obtain $K_{D(XI.17)}$ in terms of Z as expressed below

$$K_{D(XI.17)} = \frac{(1+Z)(3+Z)}{(1-Z)(3-Z)}. \qquad (XI.19)$$

We may write for an ideal orthoclase ($KAlSi_3O_8$)

$$-RT \ln K_{D(XI.17)} = \Delta G^0 \qquad (XI.20)$$

where ΔG^0 is the difference in the energy of mixing for the two standard states one in which $x_{Al\text{-}T1}$ is 0.75 and the other in which $x_{Al\text{-}T2}$ is 0.75. These site occupancies correspond to $Z = 1$ and $Z = -1$ respectively. Equations expressing ΔG^0 in terms of the energies of mixing of the feldspar in the two states have been developed by THOMPSON (1969). ΔH^0 and ΔS^0 have been computed by HOVIS (1971). Unfortunately these quantities depend on the value of the ordering parameter Z which cannot be determined precisely. From HOVIS (1971) we have ΔS^0 as -3.80 to -2.76 cal/degree/mole and ΔH^0 as -6071 to -3736 cal/mole depending on how Z is calculated from lattice parameters and assuming that the most disordered sanidine represents an equilibrium temperature of $1052°$ C. With all these assumptions, a calculation for equilibration temperature for an adularia (Spencer B, $Z = 0.56$) using Eq. (XI.20) results in temperatures between 512 ($\Delta S^0 = -3.80$) to 165°C ($\Delta S^0 = -2.76$) (see Hovis, 1971).

Although these results are very approximate, the method is definitely promising. Note that much of the uncertainty in the temperature estimation is due to the uncertainty in value of Z.

3. Other Geothermometers

Many crystalline solutions show unmixing below a certain critical temperature. This property may sometimes be used as an indicator of P and T. Plagioclases (An_2 to An_{18}) unmix into two phases with compositions in the range An_{0-1} and An_{25-28} below a certain critical temperature. This temperature has not been determined but it probably lies below 873 °K, which may be the temperature of Al–Si ordering in the plagioclase. (See Ribbe, 1960, 1962; Brown, 1960, 1962). An experimental determination of peristerite solvus could be very useful for the estimating temperature of crystallization. Crawford (1966) has made a useful study of the composition of plagioclases in rocks of varying metamorphic grade. Such petrologic studies should be useful to those attempting to fix the peristerite solvus experimentally.

Subsolidus compositional relationships between coexisting phases such as calcite-dolomite (Goldsmith and Newton, 1969), enstatite-diopside and forsterite-monticellite (Warner, 1971) may also be used for temperature estimates. It should be possible to consider unmixing in some other crystalline solutions such as magnetite (see Rumble, 1970) as probable indicators of P and T in natural assemblages.

Appendix:
Computer Programs for Calculation
of Thermodynamic Functions of Mixing
in Crystalline Solutions

As most of the important rock-forming minerals are crystalline solutions of two or more components, it is necessary that mineralogists and petrologists become more familiar with the thermodynamic behaviour of crystalline solutions. The experimental data on the silicate solutions are meagre and quantitative calculations for many important minerals are not possible at present. However, a semi-quantitative study of the data available from phase-diagrams and natural mineral assemblages may often be suitably used for a better understanding of the experimental and natural assemblages. The programs described here are useful in various calculations for the thermodynamic functions of mixing and the activity-composition relations in minerals. These programs may be particularly useful to graduate students who may want to familiarize themselves with thermodynamic behaviour of solutions by computing various real or hypothetical problems. The thermodynamic equations used here are taken from GUGGENHEIM (1952, 1967), PRIGOGINE and DEFAY (1954) and KING (1969).

Symbols and notations used here are somewhat different from those used in the rest of the book and have been explained at appropriate places. Equation numbers corresponding to those in the book are given in square brackets.

1. Program BETA

a) Purpose

This program may be used to solve the equation

$$
\ln x_{1A} + \frac{zq_1}{2} \ln\left[1 + \frac{\varphi_{2A}(\beta - 1)}{\varphi_{1A}(\beta + 1)}\right]
$$
$$
= \ln x_{1B} + \frac{zq_1}{2} \ln\left[1 + \frac{\varphi_{2B}(\beta' - 1)}{\varphi_{1B}(\beta' + 1)}\right]
$$

(1) [VI.12]

where x_{1A} and x_{1B} are the mole fractions of component 1 in A and B coexisting phases, φ_1 and φ_2 are constant fractions defined by

$$\varphi_1 = \frac{x_1 q_1}{x_1 q_1 + x_2 q_2}, \qquad \varphi_2 = \frac{x_2 q_2}{x_1 q_1 + x_2 q_2}; \qquad (2)\ [II.18]$$

and β and β' are for A and B phases, respectively, and are given by the relation

$$\beta = \{1 + 4\varphi_1 \varphi_2 [\exp(2W/zRT) - 1]\}^{1/2}; \qquad (3)\ [II.11]$$

q_1 and q_2 are contact factors and for very similar components, such as Fe^{2+} and Mg^{2+} may be taken as unity. GUGGENHEIM (1952) considered zq_1 as the number of sites which are neighbors of a molecule of type represented by component 1.

The notations x_{1A} and x_{1B} correspond to $x_{A\text{-}\alpha}$ and $x_{B\text{-}\alpha}$ and x_{2A} and x_{2B} correspond to $x_{A\text{-}\beta}$ and $x_{B\text{-}\beta}$ in [VI.12].

b) Numerical Method

Setting $y = 2W/zRT$, let

$$f(y) = \frac{zq_1}{2} \ln\left[\frac{(\beta'(y) + 1)(\beta(y) + 1 - 2\varphi_{2A})}{(\beta(y) + 1)(\beta'(y) + 1 - 2\varphi_{2B})}\right]$$
$$+ \ln\left(\frac{x_{1A}}{x_{1B}}\right) + \frac{zq_1}{2} \ln\left(\frac{\varphi_{1B}}{\varphi_{1A}}\right). \qquad (4)$$

Then the problem of finding y^* such that (1) is satisfied becomes the problem of finding y^* such that in (4) $f(y^*) = 0$.

The method of solution is of bounding the zero, y^*, above and below by y_1 and y_2 such that after the ith iteration

$$|y_1^{(i)} - y_2^{(i)}| = (y_1^{(0)} - y_2^{(0)})/2^i$$

where $y_1^{(0)}$ and $y_2^{(0)}$ are the initial bounds input to the program. Note: the assumption,

$$y_1^{(i)} \leq y^* \leq y_2^{(i)}$$

is equivalent to

$$f(y_1^{(i)}) f(y_2^{(i)}) \leq 0; \ y^{*(0)}$$

has initial value

$$(y_1^{(0)} + y_2^{(0)})/2.$$

At each iteration $f(y^{*(i)})$ is evaluated. If $|f(y^{*(i)})| < \varepsilon$ (in this program $\varepsilon = 10^{-4}$), then the zero is considered found with $y^* = y^{*(i)}$; else if

$$f(y_K^{(i)}) \ f(y^{*(i)}) \geq 0$$

then

$$y_K^{(i+1)} = y_K^{(i)}, \ y_{\text{mod}(K,\,2)+1}^{(i+1)} = y^{*(i)} \quad \text{and} \quad y^{*(i+1)} = \frac{y_1^{(i+1)} + y_2^{(i+1)}}{2}$$

for $k = 1$ or 2, provided i does not exceed a predetermined maximum, in which case the search for the zero is considered a failure.

For each set $(z, q_1, q_2, \varphi_{1A}, \varphi_{1B})$ of data the program prints the following information: z, q_1, q_2, φ_{1A}, φ_{1B}, $y^{*(i)}$, $f(y^{*(i)})$, i, where $y^* = y^{*(i)}$ if zero found else $y^{*(i)}$ is the final estimate when the search failed.

There are two cases where failure can occur:

(1) $f(y_1^{(0)}) f(y_2^{(0)}) > 0$;

that is, y_1 and y_2 did not bound y^*;

(2) $y_2^{(0)} - y_1^{(0)}$

was too large.

When either occurs additional information is printed as an aid:

(1) $y^{*(i)}, f(y^{*(i)}), \beta(y^{*(i)}), \beta'(y^{*(i)}), e^{y^{*(i)}}$

for each value i assumed.

c) Notation Used in Program BETA

X1A:	x_{1A}	X2A:	x_{2A}	
X1B:	x_{1B}	X2B:	x_{2B}	
Z:	z			
Q1:	q_1	Q2:	q_2	
P1A:	φ_{1A}	P2A:	φ_{2A}	
P1B:	φ_{1B}	P2B:	φ_{2b}	
Y1:	$y_1^{(0)}$	Y2:	$y_2^{(0)}$	Y3 $= y^{*(0)}$
Y(1):	$y_1^{(i)}$	Y(2):	$y_2^{(i)}$	Y(3) $= y^{*(i)}$
BETA:	$\beta(y)$	BETAP:	$\beta'(y)$	
F:	$f(y)$			
EY:	e^y			
X:	current value of $\beta(y)$			
XP:	current value of $\beta'(y)$			
FY:	current value of $f(y)$			
ITER:	i			

d) Input to and Output from Program BETA

Card 1: column 1–5 (right-adjusted)
NX: number of pairs of (x_{1A}, x_{1B}) to be evaluated for a given z, q_1, q_2.

Card 2: columns 1–5, 6–10, ..., 76–80 (fixed point format) (X1A, X1B) up to 8 pairs per card. Card 2 format is repeated until the NX pairs of (X1A, X1B) are entered 8 to a card, except possibly the last.

Last Card: columns 1–5, 6–10, ..., 21–25 (fixed point format) Z, Y1, Y2, Q1, Q2 respectively.

This set of cards constitutes a case. Multiple cases are permitted, each case stacked one behind the other.

Fig. 1 shows a sample set of input to program BETA while Fig. 2 shows a sample set of output.

```
  6
.005  .995  .010  .990  .025  .975  .035  .965  .067  .933  .120  .880
4.0   0.0   6.0   .95   1.05
```

Fig. 1. Sample input to program BETA

Z	X1A	X1B	Y	F(Y)	#	Q1	Q2
4.	0.005	0.995	0.55673E 01	0.38147E-05	15	0.95	1.05
4.	0.010	0.990	0.49726E 01	0.42915E-04	15	0.95	1.05
4.	0.025	0.975	0.42382E 01	-0.44823E-04	14	0.95	1.05
4.	0.035	0.965	0.39869E 01	-0.97275E-04	14	0.95	1.05
4.	0.067	0.933	0.35372E 01	0.87738E-04	14	0.95	1.05
4.	0.120	0.880	0.31832E 01	0.21935E-04	12	0.95	1.05

Fig. 2. Sample output from program BETA

e) Listing of Program BETA

```
C         PROGRAM BETA
C         P.A.COMELLA
C         CODE 641.1
C         GODDARD SPACE FLIGHT CENTER
C         GREENBELT,MARYLAND 20771
    BETA(  EY)=SQRT(1.+4.*X1A*X2A*(EY-1.))                          00000100
    BETAP(  EY)=SQRT(1.+4.*X1B*X2B*(EY-1.))                         00000200
    F(X,XP)=ALOG((XP+1.)*(X+1.-2.*P2A)/((X+1.)*(XP+1.-2.*P2B)))*Z2  00000300
  1      +     ALOG(X1A/X1B)+ALOG(P1B/P1A)  *Z2                     00000400
    COMMON STACK(6,21),Y,FY,X,XP,EY,IT,ITER                         00000500
    INTEGER*4 OUT/6/,IN/5/                                          00000600
```

```
      REAL*4 A1(50),B1(50),Y(3),EY(3),X(3),XP(3),FY(3) ,FYP(3)        00000700
100   READ(IN,1,END=1500)NX,(A1(I),B1(I),I=1,NX)                      00000800
1     FORMAT(I5        /(16F5.0))                                     00000825
      READ(IN,5)Z,Y1,Y2,Q1,Q2                                        00000810
5     FORMAT(16F5.0)                                                  00000820
      Y3=.5*(Y1+Y2)                                                   00000815
      WRITE(OUT,3)                                                    00000850
3     FORMAT('1',T4,'Z',T13,'X1A',T22,'X1B',T40,'Y',T53,'F(Y)',T63,'#',00000875
     1    T73,'Q1',T83,'Q2')                                          00000885
300   Z2=.5*Z *Q1                                                     00000950
      DO 1200 I=1,NX                                                  00001000
      X1A=A1(I)                                                       00001100
      X1B=B1(I)                                                       00001200
      X2A=1.-X1A                                                      00001300
      X2B=1.-X1B                                                      00001400
      P1A=X1A*Q1/(X1A*Q1+X2A*Q2)                                      00001410
      P2A=1.-P1A                                                      00001420
      P1B=X1B*Q1/(X1B*Q1+X2B*Q2)                                      00001430
      P2B=1.-P1B                                                      00001440
      ITER=1                                                          00001500
      IT=0                                                            00001505
      Y(1)=Y1                                                         00001525
      Y(2)=Y2                                                         00001550
      Y(3)=Y3                                                         00001575
      DO 500 J=1,2                                                    00001600
      EY(J)=EXP(Y(J)/Z2)                                              00001700
      X(J)=BETA(EY(J))                                                00001800
      XP(J)=BETAP(EY(J))                                              00001900
      FY(J)=F(X(J),XP(J))                                             00002000
      IT=IT+1                                                         00002050
      CALL SAVE(J)                                                    00002100
      IF(ABS(FY(J)).LE..0001) GO TO 1000                             00002200
500   CONTINUE                                                        00002300
600   IF(ITER.LE.20) GO TO 625                                        00002350
      WRITE(OUT,4) Z,X1A,X1B,STACK                                    00002351
4     FORMAT(F4.0,5X,F6.3,5X,F6.3,T94,'NON-CONVERGENCE'/T94,'BETA',    00002352
     1    T105,'BETA-P',T114,'EXP(2Y/(Z*Q1))',T130,'ITER'/            00002353
     2    (31X,2E13.5,30X,3E13.5,I5))                                 00002354
      GO TO 1000                                                      00002355
     1    ' X=A-INVERSE*B :LEAST SQUARES COEFFICIENTS'/3D20.8)        00029250
6     FORMAT(T30,'K',T70,'M',T90,'N',T50,'LN(K)'/T20,D20.8,20X,       00032500
     1    2D20.8/T4,'J',T15,'XAB',T35,'XAA',T55,'Y',T75,              00032700
     2    'Y-CALC',T95,'R'/(I5,5D20.8))                               00032800
625   EY(3)=EXP(Y(3)/Z2)                                              00002360
      X(3)=BETA(EY(3))                                                00002365
      XP(3)=BETAP(EY(3))                                              00002370
      FY(3)=F(X(3),XP(3))                                             00002380
      IT=IT+1                                                         00002383
      CALL SAVE(J)                                                    00002385
      J=3                                                             00002390
      IF(ABS(FY(J)).LE..0001) GO TO 1000                             00002395
      J2=2                                                            00002400
      DO 700 J1=1,2                                                   00002500
      SIGN=FY(J)*FY(J1)                                               00002600
      IF(SIGN.GT.0.) GO TO 700                                        00002700
      Y(J2)=Y(J)                                                      00002710
      FY(J2)=FY(J)                                                    00002720
      Y(J)=.5*(Y(J)+Y(J1))                                            00002730
      ITER=ITER+1                                                     00002740
      GO TO 600                                                       00002750
700   J2=1                                                            00002760
1000  WRITE(OUT,2)Z,X1A,X1B,Y(J),FY(J),ITER,Q1,Q2                     00002770
2     FORMAT(F4.0,5X,F6.3,5X,F6.3,5X,2E13.5,5X,I5,2F10.3)            00003300
1200  CONTINUE                                                        00003400
      GO TO 100                                                       00003700
1500  RETURN                                                          00003800
      END                                                             00003900
      SUBROUTINE SAVE(J)                                              00003905
      COMMON STACK(5,21),ISTACK(21),YSTACK(3,5),IT,ITER               00003910
      DO 100 I=1,5                                                    00003915
      STACK(I,IT)=YSTACK(J,I)                                         00003920
100   CONTINUE                                                        00003925
      ISTACK(IT)=ITER                                                 00003930
      RETURN                                                          00003935
      END                                                             00003940
```

2. Program REGSOL 1

a) Purpose

This program may be used to analyze the distribution of a component between two binary crystalline solutions which are now assumed to be "simple mixtures" (GUGGENHEIM, 1967).

b) Numerical Method

The relation between $x_{A\text{-}\alpha}$, $x_{A\text{-}\beta}$, W_α, W_β, and K is given by

$$\ln \frac{x_{A\text{-}\beta}(1 - x_{A\text{-}\alpha})}{(1 - x_{A\text{-}\beta})(x_{A\text{-}\alpha})} = \ln K + \frac{W_\alpha}{RT}(1 - 2x_{A\text{-}\alpha}) - \frac{W_\beta}{RT}(1 - 2x_{A\text{-}\beta}) \quad (5)\,[\text{IV.11}]$$

where x's are mole fractions of A and B in α and β, W's are "interchange" energies and K, the equilibrium constant.

Given (5) and a set of NX observations $x_{A\text{-}\alpha i}$, $x_{A\text{-}\beta i}$, $i = 1, 2, \ldots, NX$, the problem is to find the best estimates for

$$K, \frac{W_\alpha}{RT'}, \quad \text{and} \quad \frac{W_\beta}{RT'},$$

according to the method of least squares. Let

$$M = \frac{W_\alpha}{RT}, N = W_\beta/RT, X_1 = 1,$$

$$X_2 = 1 - 2x_{A\text{-}\alpha}, X_3 = 1 - 2x_{A\text{-}\beta}$$

$$Y = \ln \frac{(1 - x_{A\text{-}\alpha})}{x_{A\text{-}\alpha}} \cdot \frac{x_{A\text{-}\beta}}{(1 - x_{A\text{-}\beta})}$$

$$k' = \ln K,$$

(5) can be rewritten as

$$y = k' X_1 + N X_2 - M X_3. \quad (6)$$

The set of original observations $(x_{A\text{-}\alpha i}, x_{A\text{-}\beta i})$ are now transformed into the sequence of observations $(X_{1i}, X_{2i}, X_{3i}, Y_i)$, $i = 1, 2, \ldots, NX$ which can be used in (6) to obtain the coefficients, k', N, M, in the same way as in a program of matrix inversion.

c) Notation Used in Program REGSOL 1

XAA: $x_{A-\alpha}$
XAB: $x_{A-\beta}$
XAA2: X_2
XAB2: X_3
K0: k'
M0: M
N0: N
CAY: k
Y: y
YEST: y as calculated using the least squares coefficients, k', M, N
R: y estimate $-y$
KCALC: k $-$ calculated from (6) holding M, N constant
RK: R calculated $-k$

CHISQ: $\sum\limits_{1}^{NX} R^2/Y$

KCHI: $\sum\limits_{1}^{NX} RK^2/k$

d) Input to and Output from the Program REGSOL 1

Card 1: column 1–5 (right-adjusted) NX: numbers of pairs $(x_{A-\alpha} x_{A-\beta})$.
Card 2: columns 1–10, 11–20, ..., 71–80 $(x_{A-\alpha}, x_{A-\beta})$ up to 4 pairs per card. Card 2 format is repeated until the NX pairs $(x_{A-\alpha}, x_{A-\beta})$ are entered 4 to a card, except possibly the last.

These cards constitute a case. Multiple cases are permitted, each case stacked one behind the other.

For each case the following information is printed:
(1) The least squares matrix, A, by column.
(2) The B vector (the solution $X = A^{-1}B$).
(3) A^{-1}, X which contains the least squares coefficients.
(4) CAY, K0, M0, N0, (J, XAB(J), XAA(J), y(J), YEST(J), R(J), J = 1, NX).
(5) CHISQ.
(6) (J, CAY, KCALC(J), RK(J), J = 1, NX).
(7) KCHI.

Fig. 3 shows a sample set of input to REGSOL 1 while Fig. 4 shows a sample set of output.

```
     9
  0.021        0.341        0.070        0.692        0.094        0.815        0.136        0.864
  0.258        0.902        0.341        0.899        0.533        0.907        0.033        0.475
  0.029        0.539
```

Fig. 3. Sample input to program REGSOL 1

```
A-MATRIX (BY COLUMN) :LEAST SQUARES MATRIX
    0.90000000D 01      0.59700000D 01     -0.38680000D 01
   -0.59700000D 01     -0.49461480D 01      0.16830920D 01
   -0.38680000D 01     -0.16830920D 01      0.31298640D 01
B-VECTOR
   -0.29237103D 02     -0.20570702D 02      0.12118246D 02

A-INVERSE (BY COLUMN)
    0.21041566D 01      0.20254862D 01      0.15111833D 01
   -0.20254862D 01     -0.21972169D 01     -0.13216109D 01
    0.15111833D 01      0.13216109D 01      0.14763786D 01
=A-INVERSE*B :LEAST SQUARES COEFFICIENTS
   -0.15408793D 01      0.19945518D 01      0.89496261D 00


Y=LN(XAB*(1-XAA)/(XAA*(1-XAB)))
Y-CALC=LN(K)-M*(1-2*XAB)+N*(1-2*XAA)
R=YEST-Y
```

	K	LN(K)	M	N
	0.21419268D 00	-0.15408793D 01	0.19945518D 01	0.89496261D 00

J	XAB	XAA	Y	Y-CALC	R
1	0.21000000D-01	0.34100000D 00	-0.31831681D 01	-0.31570618D 01	0.16106388D-01
2	0.70900000D-01	0.69200000D 00	-0.35961755D 01	-0.35998594D 01	-0.20368392D-01
3	0.94000000D-01	0.81500000D 00	-0.37485768D 01	-0.37242818D 01	0.24295052D-01
4	0.13600000D 00	0.86400000D 00	-0.36973358D 01	-0.36444457D 01	0.53390018D-01
5	0.25800000D 00	0.90200000D 00	-0.32760367D 01	-0.32257923D 01	0.50244423D-01
6	0.34100000D 00	0.89900000D 00	-0.28450036D 01	-0.28393269D 01	-0.44323333D-01
7	0.53300000D 00	0.90700000D 00	-0.21453508D 01	-0.21377384D 01	0.76123585D-02
8	0.33000000D 00	0.47500000D 00	-0.32776075D 01	-0.33594250D 01	-0.81435022D-01
9	0.29000000D-01	0.53900000D 00	-0.36673482D 01	-0.34895541D 01	0.17779404D 00

```
CHISQ=SUMMATION(   R**2/Y    ) =    -0.25356160D-01


K-CALC=(EXP(M)*XAB2)/EXP(N0*XAA2))*YEXP
RK=K-CALC - KO
```

J	KO	K-CALC	RK
1	0.21419268D 00	0.21077045D 00	-0.34222366D-02
2	0.21419268D 00	0.26258108D 00	0.48388400D-01
3	0.21419268D 00	0.20951557D 00	-0.51411176D-02
4	0.21419268D 00	0.20305685D 00	-0.11135835D-01
5	0.21419268D 00	0.20369659D 00	-0.10496095D-01
6	0.21419268D 00	0.22389996D 00	0.97072738D-02
7	0.21419268D 00	0.21256836D 00	-0.16243212D-02
8	0.21419268D 00	0.23236537D 00	0.18172691D-01
9	0.21419268D 00	0.17930387D 00	-0.34888814D-01

```
CHISQ=SUMMATION(RK**2/KO)=      0.19879760D-01
```

Fig. 4. Sample output from program REGSOL 1

e) Listing of Program REGSOL 1

```
C          PROGRAM:REGSOL1
C          P.A.COMELLA
C          CODE 641.1
C          GODDARD SPACE FLIGHT CENTER
           IMPLICIT REAL*8 (A-H,O-Z)                                       00022600
           REAL*8 K(10),M(10),N(10),XAA(100),XAB(100),XAA2(100),           00022700
          1   XAAQ(100),XAB2(100),XABQ(100),XABA(100),Y(100),             00022800
          2   YEST(100)   ,KO,MO,NO,R(100),A(3,3),B(3) ,YEXP(100)          00022900
          4   ,KCALC(100),RK(100),KCHI                                     00023010
           COMMON KO,MO,NO,XAB,XAA,XAB2,XAA2,NX                            00023100
           IOUT=6                                                         00023300
           IN=5                                                           00023400
C          INPUT
200        READ(IN,2,END=1600) NX,(XAB(I),XAA(I),I=1,NX)                   00023800
2          FORMAT( I5/(8F10.3))                                           00023900
           DO 300 I=1,NX                                                  00024000
           XAA2(I)=1.D0-2.D0*XAA(I)                                       00024100
           XAAQ(I)=XAA2(I)*XAA2(I)                                        00024200
           XAB2(I)=1.D0-2.D0*XAB(I)                                       00024300
           XABQ(I)=XAB2(I)*XAB2(I)                                        00024400
           XABA(I)=XAB2(I)*XAA2(I)                                        00024500
           YEXP(I)= ((XAB(I)*(1.D0-XAA(I)))/(XAA(I)*(1.D0-XAB(I))))       00024600
           Y(I)=DLOG(YEXP(I))                                             00024700
300        CONTINUE                                                       00024800
           JGO=1                                                          00025700
           WRITE(IOUT,3)                                                  00025800
3          FORMAT('1       PROGRAM REGSOL1')                              00025900
           DO 600 IM=1,3                                                  00026500
           B(IM)=0.D0                                                     00026600
           DO 600 JM=IM,3                                                 00026700
600        A(IM,JM)=0.D0                                                  00026800
           DO 700 I=1,NX                                                  00026900
           A(1,2)=A(1,2)+XAB2(I)                                          00027000
           A(1,3)=A(1,3)+XAA2(I)                                          00027100
           A(2,2)=A(2,2)+XABQ(I)                                          00027200
           A(2,3)=A(2,3)+XABA(I)                                          00027300
           A(3,3)=A(3,3)+XAAQ(I)                                          00027400
           B(1)=B(1)+Y(I)                                                 00027500
           B(2)=B(2)+Y(I)*XAB2(I)                                         00027600
           B(3)=B(3)+XAA2(I)*Y(I)                                         00027700
700        CONTINUE                                                       00027800
           A(1,1)=NX                                                      00027900
           A(2,1)=A(1,2)                                                  00028000
           A(1,2)=-A(1,2)                                                 00028100
           A(2,2)=-A(2,2)                                                 00028200
           A(3,1)=A(1,3)                                                  00028250
           A(3,2)=-A(2,3)                                                 00028300
           WRITE(IOUT,7) A,B                                              00028800
7          FORMAT('0A-MATRIX (BY COLUMN):LEAST SQUARES MATRIX'/3D20.8/3D20.8/ 00028900
          1   3D20.8/' B-VECTOR'/3D20.8)                                  00029000
           CALL MATINV(A,3,B,1,DETERM)                                    00029100
           WRITE(IOUT,1) A,B                                              00029200
1          FORMAT('0A-INVERSE (BY COLUMN)'/3D20.8/3D20.8/3D20.8/          00029225
           KO=B(1)                                                        00029300
           MO=B(2)                                                        00029400
           NO=B(3)                                                        00029500
           CAY=DEXP(KO)                                                   00029600
           CHISQ=0.D0                                                     00029800
           KCHI=0.D0                                                      00029850
           DO 800 J=1,NX                                                  00029900
           YEST(J)   = KO-MO*XAB2(J)+NO*XAA2(J)                           00030000
           R(J)=YEST(J)   -Y(J)                                           00030100
           CHISQ=   R(J)**2/   Y(J)+CHISQ                                 00030400
           KCALC(J)=DEXP(MO*XAB2(J)-NO*XAA2(J))*YEXP(J)                   00030410
           RK(J)=KCALC(J)-CAY                                             00030420
           KCHI=KCHI+RK(J)**2/CAY                                         00030430
800        CONTINUE                                                       00030500
           WRITE(IOUT,10)                                                 00030600
10         FORMAT(////' Y=LN(XAB*(1-XAA)/(XAA*(1-XAB))'/' Y-CALC=LN(K)-M*' 00030700
          1   T18,'(1-2*XAB)+N*(1-2*XAA)'/' R=YEST-Y')                    00030800
850        WRITE(IOUT,6)                     CAY,KO,MO,NO,(J,XAB(J),       00031000
          1   XAA(J),Y(J),YEST(J),R(J),J=1,NX)                            00031100
           WRITE(IOUT,9) CHISQ                                            00031900
9          FORMAT(' CHISQ=SUMMATION(    R**2/Y    )=',D20.8)              00032000
           WRITE(IOUT,11)(J,CAY,KCALC(J),RK(J),J=1,NX)                    00032010
```

```
11       FORMAT(/' K-CALC=(EXP(MO*XAB2)/EXP(NO*XAA2))*YEXP'/       00032020
1        ' RK=K-CALC     -     KO'/                               00032030
2        T4,'J',T55,'KO',T75,'K-CALC',T115,'RK'/                  00032040
3        (I5,40X,2D20.8,20X,D20.8))                               00032050
         WRITE(IOUT,12) KCHI                                      00032060
12       FORMAT(' CHISQ=SUMMATION(RK**2/KO)=',D20.8)              00032070
1000     CONTINUE                                                 00032300
1200     CONTINUE                                                 00032400
1500     CONTINUE                                                 00033000
         GO TO  200                                               00033100
1600     RETURN                                                   00033200
         END                                                      00033300
         SUBROUTINE MATINV(A,N,B,M,DETERM)                        00009700
C        MATINV IS A VERSION OF THE SHARE SUBROUTINE OF THE SAME NAME.
         IMPLICIT REAL*8 (A-H,O-Z)                                00009800
         REAL*8 A(N,N),B(N,M),PIVOT(10)                           00009900
         INTEGER*4 IPIVOT(10),INDEX(10,2)                         00010000
         EQUIVALENCE (IROW,JROW),(ICOLUM,JCOLUM),(AMAX,T,SWAP)    00010100
         DETERM=1.DO                                              00010200
         DO 20 J=1,N                                              00010300
20       IPIVOT(J)=0                                              00010400
         DO 550 I=1,N                                             00010500
         AMAX=0.DO                                                00010600
         DO 105 J=1,N                                             00010700
         IF(IPIVOT(J).EQ.1) GO TO 105                             00010800
         DO 100 K=1,N                                             00010900
         IF(IPIVOT(K)-1) 80,100,740                               00011000
80       IF(DABS(AMAX).GE.DABS(A(J,K))) GO TO 100                 00011100
         IROW=J                                                   00011200
          ICOLUM=K                                                00011300
         AMAX=A(J,K)                                              00011400
100      CONTINUE                                                 00011500
105      CONTINUE                                                 00011600
         IPIVOT(ICOLUM)=IPIVOT(ICOLUM)+1                          00011700
         IF(IROW.EQ.ICOLUM) GO TO 260                             00011800
         DETERM=-DETERM                                           00011900
         DO 200 L=1,N                                             00012000
         SWAP=A(IROW,L)                                           00012100
         A(IROW,L)=A(ICOLUM,L)                                    00012200
200      A(ICOLUM,L)=SWAP                                         00012300
         IF(M.LE.0) GO TO 260                                     00012400
         DO 250 L=1,M                                             00012500
         SWAP=B(IROW,L)                                           00012600
         B(IROW,L)=SWAP                                           00012700
250      B(ICOLUM,L)=SWAP                                         00012800
260      INDEX(I,1)=IROW                                          00012900
         INDEX(I,2)=ICOLUM                                        00013000
         PIVOT(I)=A(ICOLUM,ICOLUM)                                00013100
         DETERM=DETERM*PIVOT(I)                                   00013200
          A(ICOLUM,ICOLUM)=1.DO                                   00013300
         DO 350 L=1,N                                             00013400
350      A(ICOLUM,L)=A(ICOLUM,L)/PIVOT(I)                         00013500
         IF(M.LE.0) GO TO 380                                     00013600
         DO 370 L=1,M                                             00013700
         B(ICOLUM,L)=B(ICOLUM,L)/PIVOT(I)                         00013800
370      CONTINUE                                                 00013900
380      DO 550 L1=1,N                                            00014000
         IF(L1.EQ.ICOLUM) GO TO 550                               00014100
         T=A(L1,ICOLUM)                                           00014200
         A(L1,ICOLUM)=0.DO                                        00014300
         DO 450 L=1,N                                             00014400
450      A(L1,L)=A(L1,L)-A(ICOLUM,L)*T                            00014500
         IF(M.LE.0) GO TO 550                                     00014600
         DO 500 L=1,M                                             00014700
500      B(L1,L)=B(L1,L)-B(ICOLUM,L)*T                            00014800
550      CONTINUE                                                 00014900
         DO 710 I=1,N                                             00015000
         L=N+1-I                                                  00015100
         IF(INDEX(L,1).EQ.INDEX(L,2)) GO TO 710                   00015200
         JROW=INDEX(L,1)                                          00015300
         JCOLUM=INDEX(L,2)                                        00015400
         DO 705 K=1,N                                             00015500
         SWAP=A(K,JROW)                                           00015600
         A(K,JROW)=A(K,JCOLUM)                                    00015700
         A(K,JCOLUM)=SWAP                                         00015800
705      CONTINUE                                                 00015900
710      CONTINUE                                                 00016000
740      RETURN                                                   00016100
         END                                                      00016200
```

3. Program REGSOL 2

a) Purpose

If the data on K, W_α/RT, W_β/RT are available, we may calculate $x_{A\text{-}\alpha}$ or $x_{A\text{-}\beta}$, (given one or the other) in (5) and plot these on a Roozeboom figure. This provides us with a distribution curve or isotherm representing the distribution of a component between two binary solutions.

b) Numerical Method

REGSOL 2 assumes that in (5) K, W_α/RT, W_β/RT and $x_{A\text{-}\beta}$ are given, the problem then being to find $x_{A\text{-}\alpha}$. To accomplish this (5) is transformed as follows:

$$g(x_{A\text{-}\alpha}) = (1 - x_{A\text{-}\alpha})\exp(-N(1 - 2x_{A\text{-}\alpha})) - f(x_{A\text{-}\beta})\,x_{A\text{-}\alpha} \qquad (7)$$

where

$$M = W_\alpha/RT, N = W_\beta/RT,$$

$$f(x_{A\text{-}\beta}) = \exp\ln\left(k\frac{(1 - x_{A\text{-}\beta})}{x_{A\text{-}\beta}}\right) - M(1 - 2x_{A\text{-}\beta}).$$

The Newton-Raphson method is then applied to (7) with

$$x_{A\text{-}\alpha(0)} = x_{A\text{-}\beta}/(x_{A\text{-}\beta} + k(1 - x_{A\text{-}\beta}))$$

the zeroth estimate of $x_{A\text{-}\alpha}$.
Each subsequent estimate is given by

$$x_{A\text{-}\alpha(i+1)} = x_{A\text{-}\alpha(i)} - \frac{g(x_{A\text{-}\alpha(i)})}{g'(x_{A\text{-}\alpha(i)})} \qquad (8)$$

where

$$g'(x_{A\text{-}\alpha}) = (2N(1 - x_{A\text{-}\alpha}) - 1) - 1\exp(-N(1 - 2x)) - f(x_{A\text{-}\beta}).$$

Whenever $[g(x_{A\text{-}\alpha(i)})] < \varepsilon$, ($\varepsilon$ is set in the program at 10^{-4}), the zero is said to have been found. This method fails in that region of the $x_{A\text{-}\alpha}$ vs. $x_{A\text{-}\beta}$ curve where the slope is parallel to the $x_{A\text{-}\alpha}$ axis.

c) Notation Used in Program REGSOL 2

XAB: $x_{A\text{-}\beta}$
XAAC: $x_{A\text{-}\alpha}$
C: $f(x_{A\text{-}\beta})$
K: K0
M: M0
N: N0
ALK: k – calc 6 using XAAC, XAB, M, N in (6)
R: K0 – ALK

d) Input to and Output from Program REGSOL 2

Card 1: column 1–10, 11–20, 21–30 (fixed point format) K0, M0, N0, respectively.

Card 2: column 1–5 (right-adjusted) NX: number of observations $x_{A\text{-}\beta}$ for which $x_{A\text{-}\alpha}$ is to be found.

Card 3: columns 1–10, 11–20, ..., 71–80 (fixed point format) $x_{A\text{-}\beta}$ up to 8 observations per card. Card 2 format is repeated until the NX values of $x_{A\text{-}\beta}$ are entered 8 to a card, except possibly the last card.

This set of cards constitutes a case. Multiple cases are permitted, each case stacked one behind the other.

For each observation the following output is provided if the zero has been found:

$$x_{A\text{-}\beta}, \, x_{A\text{-}\alpha}, \, k_0, \, k\text{-calculated}, \, R.$$

For the entire case a graph of $x_{A\text{-}\alpha}$ versus $x_{A\text{-}\beta}$ is given.

When the Newton-Raphson method fails to find the zero the following information is given:

$$x_{A\text{-}\beta}, \, x_{A\text{-}\alpha(i)}, \, f\left(x_{A\text{-}\alpha(i)}\right)$$

for each iteration i.

```
0.2823      2.0000      1.3390
   16
0.03        0.05        0.06        0.08        0.10        0.15        0.20        0.25
0.30        0.40        0.50        0.60        0.70        0.80        0.90        0.95
```

Fig. 5. Sample input to program REGSOL 2

```
FOR XAB=   0.25000E 00   ZERO WAS NOT FOUND.TRACE' OF ITERATIONS FOLLOWS:
            X                F(X)
0.54144782E 00    0.34369034E 00
0.65933743E 01   -0.68313744E 08
0.62433300E 01   -0.25080240E 08
0.58947420E 01   -0.92051840E 07
0.55477962E 01   -0.33774710E 07
0.52027168E 01   -0.12387390E 07
0.48597736E 01   -0.45410481E 06
0.45192976E 01   -0.16636613E 06
0.41817017E 01   -0.60902648E 05
0.38475027E 01   -0.22272457E 05
0.35173635E 01   -0.81342930E 04
0.31921453E 01   -0.29654358E 04
0.28729849E 01   -0.10783801E 04
0.25614128E 01   -0.39076123E 03
0.22595224E 01   -0.14085666E 03
0.19702377E 01   -0.50367310E 02
0.16977329E 01   -0.17775421E 02
0.14481325E 01   -0.61282244E 01
0.12306871E 01   -0.20158863E 01
0.10592642E 01   -0.59501708E 00
```

Fig. 6. Sample output from program REGSOL 2 trace when zero not found

Fig. 5 shows a sample set of input to program REGSOL 2 while Fig. 6 and 7 show a sample set of output.

```
        K= 0.2823E 00      M= 0.2000E 01      N= 0.1339E 01
NEWTON-RAPHSON METHOD: ITERATION #  3
  J         X AB        XAA      XAA(CALC)       K(CALC)        K(INPUT)       K-K(CALC)
  1     0.3000E-01               0.2908E 00    0.2823E 00    0.2823E 00    0.1311E-05
  2     0.5000E-01               0.5882E 00    0.2823E 00    0.2823E 00   -0.2980E-06
  3     0.6000E-01               0.6809E 00    0.2822E 00    0.2823E 00    0.5084E-04
  4     0.8000E-01               0.7757E 00    0.2823E 00    0.2823E 00    0.1848E-05
  5     0.1000E 00               0.8220E 00    0.2823E 00    0.2823E 00    0.4292E-05
  6     0.1500E 00               XXXX    ZERO NOT FOUND
  7     0.2000E 00               XXXX    ZERO NOT FOUND
  8     0.2500E 00               XXXX    ZERO NOT FOUND
  9     0.3000E 00               0.9102E 00    0.2822E 00    0.2823E 00    0.8500E-04
 10     0.4000E 00               0.9145E 00    0.2823E 00    0.2823E 00    0.4548E-04
 11     0.5000E 00               0.9150E 00    0.2823E 00    0.2823E 00    0.1848E-05
 12     0.6000E 00               0.9155E 00    0.2823E 00    0.2823E 00    0.6139E-05
 13     0.7000E 00               0.9195E 00    0.2823E 00    0.2823E 00    0.4089E-04
 14     0.8000E 00               0.9313E 00    0.2822E 00    0.2823E 00    0.7963E-04
 15     0.9000E 00               0.9562E 00    0.2823E 00    0.2823E 00    0.3219E-05
 16     0.9500E 00               0.9755E 00    0.2823E 00    0.2823E 00    0.1848E-05
```

Fig. 7. Sample output from program REGSOL 2

e) Listing of Program REGSOL 2

```
C          PROGRAM REGSOL2
C          P.A.COMELLA
C          CODE 641.1
C          GODDARD SPACE FLIGHT CENTER
C          GREENBELT,MARYLAND 20771
C
      REAL*4 K,M,N                                           00016800
      COMMON K,M,N,X(100)   ,NX                              00016900
C          INPUT
100   READ(5,2,END=1000) K,M,N                               00017000
2     FORMAT(3F10.3)                                         00017100
      READ(5,3) NX,(X(I),I=1,NX)                             00017200
3     FORMAT(I5/(8F10.3))                                    00017300
      CALL ESTMTE                                            00017400
      WRITE(6,1)                                             00017500
1     FORMAT('1       PROGRAM REGSOL2')
      GO TO 100                                              00017700
1000  RETURN                                                 00017800
      END                                                    00017900
      SUBROUTINE ESTMTE                                      00018000
      IMPLICIT REAL*4 (A-H,K-Z)                              00018100
      INTEGER*4 ITER /20/,IOUT/6/,FLAG                       00018200
1     ,NSCALE(5)/5*0/,NHL/8/,NSBH/6/,NVL/8/,NSBV/10/         00018290
      REAL*4 XAAC(100),R(100),TOL/1.D-4/ ,STACK(2,20)        00018300
      COMMON KO,MO,NO ,XAB(100)      ,JX                     00018400
      REAL*4 XAB2(100),RESID/0./ ,XALF/'XXXX'/               00018500
      DIMENSION XXAB(100),ALK(100)                           00018550
      LOGICAL*1 GRID(5200)                                   00018600
      FUN(K,M,X,X2)= EXP(ALOG(K*((1.00-X)/X))-M*X2)          00018700
      GFUN(N,X)=(1.00-X)* EXP(-N*(1.00-2.00*X))-C*X          00018800
      GPF(N,X)=(2.00*N*(1.00-X)-1.00)* EXP(-N*(1.00-2.00*X))-C  00018900
      DO 500 J=1,JX                                          00019000
      XAB2(J)=1.00-2.00*XAB(J)                               00019300
      C=FUN(KO,MO,XAB(J),XAB2(J))                            00019400
      X=XAB(J)/(XAB(J)+KO*(1.-XAB(J)))                       00019500
      DO 100 JTER=1,ITER                                     00019600
      GX=GFUN(NO,X)                                          00019700
      STACK(1,JTER)=X                                        00019800
      STACK(2,JTER)=GX                                       00019900
      IF(ABS(GX).LT.TOL) GO TO 200                           00020000
      X=X-GX/GPF(NO,X)                                       00020100
100   CONTINUE                                               00020200
      XAAC(J)=XALF                                           00020300
      WRITE(IOUT,4)XAB(J),STACK                              00020310
4     FORMAT('0FOR XAB=',E13.5,'  ZERO WAS NOT FOUND.TRACE'  00020320
1        ' OF ITERATIONS FOLLOWS:'/T12,'X',T28,'F(X)'/(2E16.8))  00020330
      GO TO 500                                              00020340
```

```
200     XAAC(J)=X                                                          00020400
        ALK(J)=ALOG((1.-X) *XAB(J)/((1.-XAB(J))*X))+MO*XAB2(J)             00020405
    1     -NO*(1.-2.*X)                                                    00020410
        ALK(J)=EXP(ALK(J))                                                 00020415
        R(J)=KO-ALK(J)                                                     00020420
500     CONTINUE                                                           00020500
        RESID=0.                                                           00020600
        WRITE(IOUT,1)KO,MO,NO,JTER                                         00020700
1       FORMAT('1                 PROGRAM REGSOL2'/                        00020840
    1        T10,'K=',E12.4,T30,'M=',E12.4,T50,'N=',E12.4/                 00020900
    1    ' NEWTON-RAPHSON METHOD:ITERATION #',I3/                         00021000
    2           '    J',T15,'XAB',T30,'XAA',T40,'XAA(CALC)',T55,          00021010
    3'K(CALC)',T70,'K(INPUT)',T85,'K-K(CALC)')                            00021020
        DO 700 J=1,JX                                                      00020750
        IF (XAAC(J).EQ.XALF) GO TO 600                                     00020760
        WRITE(IOUT,7)J,XAB(J),XAAC(J),ALK(J),KO,R(J)                       00020770
7       FORMAT(I5,E15.4,E30.4,3E15.4)                                      00020780
        GO TO 700                                                          00020790
600     WRITE(IOUT,8)J,XAB(J),XALF                                         00020800
8       FORMAT(I5,E15.4,20X,A4,'     ZERO NOT FOUND')                      00020810
        XAAC(J)=-1.0                                                       00020820
700     CONTINUE                                                           00020830
        WRITE(IOUT,3)                                                      00021300
3       FORMAT('1')                                                        00021400
        CALL PLOT1(NSCALE,NHL,NSBH,NVL,NSBV)                               00021450
        CALL PLOT2(GRID,1.,0.,1.,0.)                                       00021500
        CALL PLOT3('*',XAB,XAAC,JX)                                        00021600
        CALL PLOT4(8,          XAA')                                       00021700
        WRITE(IOUT,2)                                                      00021800
2       FORMAT(/T25,'XAB')                                                 00021900
        WRITE(IOUT,5) KO,MO,NO                                             00022000
5       FORMAT(' PROGRAM REGSOL2'/' KO=',E13.5,10X,'MO=',E13.5,10X,'NO=',  00022100
    1      E13.5)                                                          
        RETURN                                                             00022200
        END                                                                00022300
```

4. Program QUASI

a) Purpose

If we have data on a complete distribution isotherm, this program may be used to find $2W/zRT$ for each of the two crystalline solutions assuming that they are regular solutions with the quasi-chemical approximation.

b) Numerical Method

The basic equation is

$$K = K_{\mathrm{D}} \cdot \frac{\left\{1 + \dfrac{\varphi_{1\mathrm{A}}(\beta - 1)}{\varphi_{2\mathrm{A}}(\beta + 1)}\right\}^{\frac{z_1 q_2}{2}} \left\{1 + \dfrac{\varphi_{2\mathrm{B}}(\beta' - 1)}{\varphi_{1\mathrm{B}}(\beta' + 1)}\right\}^{\frac{z_2 q_1}{2}}}{\left\{1 + \dfrac{\varphi_{2\mathrm{A}}(\beta - 1)}{\varphi_{1\mathrm{A}}(\beta + 1)}\right\}^{\frac{z_1 q_1}{2}} \left\{1 + \dfrac{\varphi_{1\mathrm{B}}(\beta' - 1)}{\varphi_{2\mathrm{B}}(\beta' + 1)}\right\}^{\frac{z_2 q_2}{2}}} \quad (9) \,[\mathrm{V.7}]$$

where symbols 1 and 2 correspond to components B and A in Eqs. [V.7] (VI.12 and VI.13) and symbols A and B in the present equation correspond to phases α and β in Eqs. (VI.12 and VI.13). The symbols z_1 and z_2

correspond to z_α and z_β in the Eqs. (VI.12 and VI.13). Letting

$$f_{1A} = \left[1 + \frac{\varphi_{2A}(\beta - 1)}{\varphi_{1A}(\beta - 1)}\right]^{\frac{z_1 q_1}{2}},$$

$$f_{2A} = \left[1 + \frac{\varphi_{1A}(\beta - 1)}{\varphi_{2A}(\beta + 1)}\right]^{\frac{z_1 q_2}{2}},$$

$$f_{1B} = \left[1 + \frac{\varphi_{2B}(\beta' - 1)}{\varphi_{1B}(\beta' + 1)}\right]^{\frac{z_2 q_1}{2}},$$

$$f_{2B} = \left[1 + \frac{\varphi_{1B}(\beta' - 1)}{\varphi_{2B}(\beta' + 1)}\right]^{\frac{z_2 q_2}{2}}.$$

Eq. (9) may be written as

$$K = \frac{x_{1A} x_{2B} f_{1A} f_{2B}}{x_{2A} x_{1B} f_{2A} f_{1B}}, \tag{10}$$

which is the notation used in the program. Now

$$\beta = \left\{1 + 4x_{1A} x_{2A}\left(\exp\frac{2W}{zRT} - 1\right)\right\}^{1/2}$$

$$\beta' = \left\{1 + 4x_{1B} x_{2B}\left(\exp\frac{2W'}{zRT} - 1\right)\right\}^{1/2}. \tag{11}$$

The problem is to find

$$\frac{2W}{zRT}, \frac{2W'}{zRT}.$$

This is done by the method of non-linear least squares as outlined in the following paragraphs: Write (11)

$$f(x_{1A}, x_{1B}, k', y, y') = k' \frac{x_{2B} f_{1A} f_{2B}}{x_{1B} f_{2A} f_{1B}} \tag{12}$$

where

$$k' = \frac{1}{K}, \quad y = \frac{2W}{zRT}, \quad y' = \frac{2W'}{zRT}.$$

Set

$$f_{obs} = x_{2A}/x_{1A},$$

$$V = f(x_{1A}, x_{1B}, k', y, y') - f_{obs}. \tag{13}$$

Let

$$k' = k'_{(0)} + \Delta k'$$

$$y = y_{(0)} + \Delta y$$

$$y' = y'_{(0)} + \Delta y'$$

where $k'_{(0)}, y_{(0)}, y'_{(0)}$ are initial estimates of k', y, y'.

Then linearize (13) by doing a Taylor Series expansion about

$$(k'_{(0)}, y_{(0)}, y'_{(0)}) = (0)$$

so obtaining

$$v + f_{obs} = f(x_{1A}, x_{1B}, k'_{(0)}, y_{(0)}, y'_{(0)})$$

$$+ \delta k' \frac{\delta f}{\delta k'}\bigg|_{(0)} + \delta y \frac{\delta f}{\delta y}\bigg|_{(0)} + \delta y' \frac{\delta f}{\delta y'}\bigg|_{(0)} \tag{14}$$

and now use method of least squares to solve this equation, iterating until

$$|\delta k'|, |\delta y|, |\delta y'|$$

are less than some prescribed ε.

c) Notation Used in Program QUASI

z_1:	Z1				
z_2:	Z2				
q_1:	Q1	q_2:	Q2		
q'_1:	Q1P	q'_2:	Q2P		
$k_{(0)}$:	KO	$y_{(0)}$:	YO	$y'_{(0)}$:	YPO

Number of observations of the pairs: (x_{1A}, x_{1B}): NX

x_{1A}:	X1A	x_{2A}:	X2A
φ_{1A}:	PHI1A	φ_{2A}:	PHI2A
φ_{1B}:	PHI1B	φ_{2B}:	PHI2B
β:	BETA	β':	BETAP
$\dfrac{d\beta}{dy}$:	DBDY	$\dfrac{d\beta}{dy'}$:	DBDYP
f_{1A}:	F1A	f_{2A}:	F2A
f_{1B}:	F1B	f_{2B}:	F2B
$\dfrac{df_{1A}}{dy}$:	DF1ADY	$\dfrac{df_{2A}}{dy}$:	DF2ADY
$\dfrac{f_{1A}}{f_{2A}}$:	FA12		
$\dfrac{df_{1B}}{dy'}$:	DF1BDY	$\dfrac{df_{2B}}{dy'}$:	DF2BDY
$\dfrac{f_{2B}}{f_{1B}}$:	FB12		
$\dfrac{\partial f}{\partial k'}$:	DFDK		

$$\frac{\partial f}{\partial y}: \quad \text{DFDY}$$

$$\frac{\partial f}{\partial y'}: \quad \text{DFDYP}$$

$f(x_{1A}, x_{1B}, k', y, y')$: F
f_{obs}:   Y

Determinant of Least Square Matrix: DET

Least square coefficients

$\delta k'$: A1 $\qquad\qquad$ δy: A2 $\qquad\qquad$ $\delta y'$: A3

Solution: k':   KPO
$\qquad\qquad$ k: KP
$\qquad\qquad$ y: YOO
$\qquad\qquad$ y':   YPOO

d) Input to and Output from Program QUASI

Input:

Card 1: $\qquad$ columns 1–5, 6–10, …, 41–45 Z1, Z2, Q1, Q2, Q1P, Q2P, KO, YO, YPO, respectively (fixed point format).

Card 2: $\qquad$ columns 1–5 NX.

Cards 3 & ff: columns 1–5, …, 76–80 $(x_{1A}^{(I)}, x_{1B}^{(I)})$, $I = 1, 2, …, NX$ fixed point format, 8 pairs (x_{1A}, x_{1B}) per card until NX pairs entered with a maximum of 50 pairs.

Output from program QUASI as follows:

(1) The input.

(2) for each iteration: KP, KPO, YOO, YPOO, DET, A1, A2, A3.

(3) for the final iteration

$\qquad$ x_{1A} (input), x_{1A} (calculated), $f_{1A}, f_{2A}, \beta(y)$,

$\qquad$ x_{1B} (input), x_{1B} (calculated), $f_{1B}, f_{2B}, \beta'(y)$,

$\qquad$ RESID: Y-F.

Fig. 8 shows a sample set of input while Fig. 9 shows a sampling of output.

2.0   2.0   1.0   1.0   1.0   1.0   0.15 1.5   1.0
    6
0.0210.3410.0700.6920.1110.7990.1500.8500.2620.8970.3370.902

Fig. 8. Sample input data for program QUASI

```
CONSTANTS AND INITIAL VALUES
Z1= 0.20000E 01   Z2= 0.20000E 01   Q1= 0.10000E 01   Q2= 0.10000E 01   Q1P= 0.10000E 01   Q2P= 0.10000E 01
K0= 0.15000E 00   Y0= 0.15000E 01   YP0= 0.10000E 01

        X1A      X1B
  1    0.021    0.341
  2    0.070    0.692
  3    0.111    0.799
  4    0.150    0.850
  5    0.262    0.897
  6    0.337    0.902

ITERATION: 0   K0= 0.15000E 00                 1/K0= 0.66667E 01                 Y= 0.15000E 01                 YP= 0.10000E 01

ITERATION: 1   K0= 0.10841E 00   1/K0= 0.92241E 01   Y= 0.11791E 01   YP= 0.61774E 00
               DET=               A1= 0.25574E 00   A2= -0.32086E 00   A3= -0.38226E 00

ITERATION: 2   K0= 0.10761E 00   1/K0= 0.92928E 01   Y= 0.12412E 01   YP= 0.70080E 00
               DET= 0.19199E 04   A1= 0.68787E-01   A2= 0.62034E-01   A3= 0.83056E-01

ITERATION: 3   K0= 0.10806E 00   1/K0= 0.92537E 01   Y= 0.12450E 01   YP= 0.70117E 00
               DET= 0.23756E 04   A1= -0.39152E-01  A2= 0.37947E-02   A3= 0.37367E-03

ITERATION: 4   K0= 0.10803E 00   1/K0= 0.92563E 01   Y= 0.12446E 01   YP= 0.70100E 00
               DET= 0.23633E 04   A1= 0.25766E-02   A2= -0.35054E-03  A3= -0.17069E-03

ITERATION: 5   K0= 0.10804E 00   1/K0= 0.92561E 01   Y= 0.12446E 01   YP= 0.70101E 00
               DET= 0.23615E 04   A1= -0.12327E-03  A2= 0.17373E-04   A3= 0.77288E-05

ITERATION: 6   K0= 0.10804E 00   1/K0= 0.92562E 01   Y= 0.12446E 01   YP= 0.70101E 00
               DET= 0.23569E 04   A1= 0.81758E-05   A2= 0.65433E-06   A3= -0.25030E-05

FINAL RESULTS
X1A-INPUT     X1A-CALC      F1A           F2A           BETA(Y)       X1B-INPUT     F1B           F2B           BETAP(YP)     RESID
0.21000E-01   0.21011E-01   0.31550E 00   0.10010E 01   0.10969E 01   0.34100E 00   0.13107E 01   0.10832E 01   0.13831E 01   -0.10975E-04
0.70000E-01   0.68511E-01   0.26420E 00   0.10093E 01   0.12820E 01   0.69200E 00   0.10689E 01   0.13476E 01   0.13660E 01    0.14887E-02
0.11100E 00   0.11393E 00   0.23503E 00   0.10210E 01   0.14056E 01   0.79900E 00   0.10314E 01   0.14966E 01   0.12855E 01   -0.29345E-02
0.15000E 00   0.15956E 00   0.21397E 00   0.10355E 01   0.15035E 01   0.85000E 00   0.10183E 01   0.15892E 01   0.12321E 01   -0.95568E-02
0.26200E 00   0.26104E 00   0.17352E 00   0.10927E 01   0.17064E 01   0.89700E 00   0.10091E 01   0.16925E 01   0.11728E 01    0.96017E-03
0.33700E 00   0.30166E 00   0.15578E 00   0.11441E 01   0.17914E 01   0.90200E 00   0.10083E 01   0.17047E 01   0.11658E 01    0.35336E-01
```

Fig. 9. Quasi-chemical approximation: Eq. (5.6)

e) Listing of Program QUASI

```
C
C          PROGRAM QUASI
C          QUASI CHEMICAL APPROXIMATION
C          P.A. COMELLA
C          CODE 641.1
C          GODDARD SPACE FLIGHT CENTER
C          GREENBELT,MARYLAND 20771
C
       REAL*4 KO,KP,KPO,X1A(50),X1B(50),PHI1B(50),PHI2B(50),X2B(50),    00000100
      1      X21B(50),X1AC(50),FA1(50),FA2(50),FB1(50),FB2(50),BET(50),  00000200
      2      BETP(50),FUN(50),DIFF(50),RESID(50)                         00000300
      3      ,XB12(50)                                                   00000325
       LOGICAL*1 TEST                                                    00000350
       INTEGER*4 IN/5/,OUT/6/                                            00000375
       COMMON SUM1,SUM2,SUM3,SUM4,SUM5,SUM6,SUM7,SUM8,SUM9               00000400
50     READ(IN,1,END=2000)Z1,Z2,Q1,Q2,Q1P,Q2P,KO,YO,YPO,NX,(X1A(I),X1B(I)00000500
      1      ,I=1,NX)                                                    00000600
1      FORMAT( 9F5.0/I5/(16F5.0))                                       00000700
       WRITE(OUT,3) Z1,Z2,Q1,Q2,Q1P,Q2P,KO,YO,YPO,(I,X1A(I),            00000710
      1      X1B(I),I=1,NX)                                              00000720
3      FORMAT('1',10X,'QUASI-CHEMICAL APPROXIMATION:EQUATION(5.6)'/      00000730
      1      'OCONSTANTS AND INITIAL VALUES'/                           00000740
      2      ' Z1=',E13.5,5X,'Z2=',E13.5,5X,'Q1=',E13.5,5X,'Q2=',        00000750
      3      E13.5,5X,'Q1P=',E13.5,5X,'Q2P=',E13.5/' KO=',E13.5,5X,'YO=', 00000760
      3      E13.5,4X,'YPO=',E13.5/'0    I',15X,                         00000765
      4      'X1A',17X,'X1B'/(I5,10X,F10.3,10X,F10.3))                   00000770
       Z12=0.5*Z1                                                       00000800
       Z22=0.5*Z2                                                       00000850
       ZQ1=Z12*Q1                                                       00000900
       ZQ2=Z22*Q2                                                       00001000
       ZQP1=Z12*Q1P                                                     00001100
       ZQP2=Z22*Q2P                                                     00001200
       ZQ11=ZQ1-1.0                                                     00001300
       ZQ21=ZQ2-1.0                                                     00001400
       ZQP11=ZQP1-1.0                                                   00001500
       ZQP21=ZQP2-1.0                                                   00001600
       KP=1.0/KO                                                        00001700
       KPO=KP                                                           00001800
       YOO=YO                                                           00001900
       YPOO=YPO                                                         00002000
       RSUM=0.0                                                         00002025
       TEST=.FALSE.                                                     00002050
       INDEX=0                                                          00002075
       WRITE(OUT,5)                                                     00002080
5      FORMAT('-')                                                      00002085
       WRITE(OUT,4) INDEX,KO,KPO,YOO,YPOO                               00002090
       INDEX=1                                                          00002095
       DO 200 I=1,NX                                                    00002100
       X2B(I)=1.0-X1B(I)                                                00002200
       PHI1B(I)=X1B(I)*Q1P/(X1B(I)*Q1P+X2B(I)*Q2P)                      00002300
       PHI2B(I)=1.0-PHI1B(I)                                            00002400
       X21B(I)=X2B(I)/X1B(I)                                            00002500
       X1AC(I)=X1A(I)                                                   00002600
       XB12(I)=2.0*X1B(I)*X2B(I)                                        00002650
200    CONTINUE                                                         00002700
900    SUM1=0.0                                                         00003000
       SUM2=0.0                                                         00003100
       SUM3=0.0                                                         00003200
       SUM4=0.0                                                         00003300
       SUM5=0.0                                                         00003400
       SUM6=0.0                                                         00003500
       SUM7=0.0                                                         00003600
       SUM8=0.0                                                         00003700
       SUM9=0.0                                                         00003800
       EY=EXP(YOO)                                                      00003810
       EYP=EXP(YPOO)                                                    00003820
       EY1=EY-1.0                                                       00003830
       EYP1=EYP-1.0                                                     00003840
       DO 1000 I=1,NX                                                   00013900
       XA1=X1A(I)                                                       00014000
       XA2=1.0-XA1                                                      00014100
       XA12=2.0*XA1*XA2                                                 00014200
       BETA=SQRT(1.0+2.0*XA12*EY1)                                      00014400
       BP1=BETA+1.0                                                     00014500
       BM1=BETA-1.0                                                     00014600
       BETAP=SQRT(1.0+2.0*XB12(I)*EYP1)                                 00014700
```

```
      PHI1A=XA1*Q1/(XA1*Q1+XA2*Q2)                                      00014725
      PHI2A=1.0-PHI1A                                                   00014750
      BPP1=BETAP+1.0                                                    00014800
      BPM1=BETAP-1.0                                                    00014900
      BPP1Q=BPP1*BPP1                                                   00015000
      BP1Q=BP1*BP1                                                      00015100
      DBDY=XA12*EY/BETA                                                 00015300
      DBDYP=XB12(I)*EYP/BETAP                                           00015400
      F1A=1.0+PHI2A*BM1/(PHI1A*BP1)                                     00015500
      F2A=1.0+PHI1A*BM1/(PHI2A*BP1)                                     00015600
      F1B=1.0+PHI2B(I)*BPM1/(PHI1B(I)*BPP1)                             00015700
      F2B=1.0+PHI1B(I)*BPM1/(PHI2B(I)*BPP1)                             00015800
      DF1ADY=ZQ1*F1A**ZQ11*2.0*PHI2A/(PHI1A*BP1Q)*DBDY                  00016100
      DF2ADY=ZQ2*F2A**ZQ21*2.0*PHI1A/(PHI2A*BP1Q)*DBDY                  00016200
      F1A=F1A**ZQ1                                                      00016300
      F2A=F2A**ZQ1                                                      00016400
      FA12=F1A/F2A                                                      00016500
      DF1BDY=ZQP1*F1B**ZQP11*PHI2B(I)/(PHI1B(I)*BPP1Q)*DBDYP            00016600
     1*2.0                                                              00016650
      DF2BDY=ZQP2*F2B**ZQP21*PHI1B(I)/(PHI2B(I)*BPP1Q)*DBDYP            00016700
     1*2.0                                                              00016715
      F1B=F1B**ZQP1                                                     00016725
      F2B=F2B**ZQP2                                                     00016750
      FB12=F2B/F1B                                                      00016800
      DFDK=X21B(I)*FA12*FB12                                            00016900
      DFDY=KPO*X21B(I)*FB12*(F2A*DF1ADY-F1A*DF2ADY)/(F2A*F2A)           00017000
      DFDYP=KPO*X21B(I)*FA12*(F1B*DF2BDY-F2B*DF1BDY)/(F1B*F1B)          00017100
      F=KPO*DFDK                                                        00017200
      Y=XA2/X1A (I)                                                     00017225
      YF=Y-F                                                            00017250
      X1AC(I)=1.0/(1.0+F)                                               00017300
      IF (TEST .EQ. .FALSE.) GO TO 950                                  00017400
      FA1(I)=F1A                                                        00017500
      FA2(I)=F2A                                                        00017600
      FB1(I)=F1B                                                        00017700
      FB2(I)=F12B                                                       00017800
      FB2(I)=F2B                                                        00017900
      BET(I)=BETA                                                       00018000
      BETP(I)=BETAP                                                     00018100
      FUN(I)=F                                                          00018200
      DIFF(I)=YF                                                        00018300
      RESID(I)=X1A(I)-X1AC(I)                                           00018400
      RSUM=RSUM+RESID(I)*RESID(I)                                       00018500
C     GO TO 1000                                                       00018600
950   CONTINUE                                                         00018700
      SUM1=SUM1+DFDK*DFDK                                              00019000
      SUM2=SUM2+DFDK*DFDY                                              00019100
      SUM3=SUM3+DFDK*DFDYP                                             00019200
      SUM4=SUM4+DFDK*YF                                                00019300
      SUM5=SUM5+DFDY*DFDY                                              00019400
      SUM6=SUM6+DFDY*DFDYP                                             00019500
      SUM7=SUM7+DFDY*YF                                                00019600
      SUM8=SUM8+DFDYP*DFDYP                                            00019700
      SUM9=SUM9+DFDYP*YF                                               00019800
1000  CONTINUE                                                         00019900
      IF (TEST .EQ. .TRUE.) GO TO 1100                                 00019950
      DET=DETERM(1,5,8,6,6)+DETERM(2,6,3,2,8)+DETERM(3,2,6,5,3)        00020000
      A1=(DETERM(4,5,8,6,6)+DETERM(7,6,3,2,8)+DETERM(9,2,6,5,3))/DET   00020100
      A2=(DETERM(1,7,8,6,9)+DETERM(2,9,3,4,8)+DETERM(3,4,6,7,3))/DET   00020200
      A3=(DETERM(1,5,9,6,7)+DETERM(2,4,6,2,9)+DETERM(3,2,7,5,4))/DET   00020300
      TEST=(ABS(A1/KPO).LT.1.E-05).AND.(ABS(A2/YOO).LT.1.E-05).AND.    00020400
     1     (ABS(A3/YPOO).LT.1.E-05).OR.INDEX.GT.10                     00020500
      KPO=KPO+A1                                                       00020700
      YOO=YOO+A2                                                       00020800
      YPOO=YPOO+A3                                                     00020900
      KP=1.0/KPO                                                       00021000
      WRITE(OUT,4) INDEX,KP,KPO,                                       00021015
     1     YOO,YPOO,DET,A1,A2,A3                                       00021025
      IF(TEST.EQ..TRUE.) GO TO 900                                     00021040
      INDEX=INDEX+1                                                    00021050
4     FORMAT(' ITERATION:',I3,5X,'KO=',E13.5,5X,'1/KO=',E13.5,         00021075
     1     5X,'Y=',E13.5,5X,'YP=',E13.5/18X,'DET=',E13.5,5X,           00021078
     2     ' A1=',E13.5,4X,'A2=',E13.5,5X,'A3=',E13.5)                 00021081
      GO TO 900                                                        00021100
1100  KP=1.0/KPO                                                       00021200
      WRITE(OUT,6) (X1A(I),X1AC(I),FA1(I),FA2(I),BET(I),              00021300
     1     X1B(I),FB1(I),FB2(I),BETP(I),RESID(I),I=1,NX)               00021400
6     FORMAT('-FINAL RESULTS'/' X1A-INPUT',4X,'X1A-CALC',8X,          00021500
```

```
     1        'F1A',8X,'F2A',6X,'BETA(Y)',3X,'X1B-INPUT',7X,          00021600
     2        'F1B',8X,'F2B',5X,'BETAP(YP)',2X,'RESID'/               00021625
     3        (10E13.5))                                              00021650
          GO TO 50                                                    00021700
2000      RETURN                                                      00021800
          END                                                         00021900
          REAL FUNCTION DETERM*4(I,J,K,L,M)
          COMMON SUM(20)
          DETERM=SUM(I)*(SUM(J)*SUM(K)-SUM(L)*SUM(M))
          RETURN
          END
```

5. Program TERNGAP

a) Purpose

Program TERNGAP calculates ternary miscibility gap. The input data required are the Eij's for the three binary systems (see Eqs. II.40, III.31–III.33). The program is written on the same lines as those by KAUFMAN and BERNSTEIN's (1970, MIGAP). However, TERNGAP does not work for systems which do not show miscibility gaps in at least one of the three binary systems.

b) Numerical Method

The details of the thermodynamic equations are discussed in Chapter III. The numerical methods employed are similar to KAUFMAN and BERNSTEIN's (Newton-Raphson iteration technique).

c) Notations Used in Program TERNGAP

EIJ: E_{ij}
X, Y, Z: x_1, x_2, x_3

d) Input to and Output from Program TERNGAP

Card 1: Column 1–10, 11–20, 21–30 etc. T, E_{12}, E_{21}, E_{13}, E_{31}, E_{23}, E_{32}, C. The E_{ij}'s are arranged so that the binary system 1–2 has the largest values of E_{ij}'s. Next lower set of values are assigned to the system 1–3 and similarly for the system 2–3. An example of the output is shown in Fig. 10 for a hypothetical ternary system.

EXITED GAP - DATA   T=   573. E12=   3000. E21=   6000.
  E13=   2950.   E31=   4230.   E23=   500.   E32=   200.   C=   0.

EDGE 1 SOLUTION -   0.265   0.942

   TIE LINE COORDINATES

   X1   Y1   Z1   X2   Y2   Z2

0.0000   0.7347   0.2653   0.0000   0.0583   0.9417
0.0100   0.7245   0.2655   0.0023   0.0583   0.9394
0.0200   0.7142   0.2658   0.0046   0.0583   0.9371
0.0300   0.7039   0.2661   0.0070   0.0582   0.9348
0.0400   0.6936   0.2664   0.0093   0.0582   0.9325
0.0500   0.6833   0.2667   0.0116   0.0581   0.9303
0.0600   0.6730   0.2670   0.0140   0.0580   0.9280
0.0700   0.6626   0.2674   0.0163   0.0580   0.9257
0.0800   0.6522   0.2678   0.0187   0.0579   0.9235
0.0900   0.6417   0.2683   0.0210   0.0578   0.9212
0.1000   0.6313   0.2687   0.0234   0.0577   0.9189
0.1100   0.6208   0.2692   0.0258   0.0576   0.9166
0.1200   0.6102   0.2698   0.0282   0.0575   0.9143
0.1300   0.5997   0.2703   0.0306   0.0573   0.9121
0.1400   0.5890   0.2710   0.0331   0.0572   0.9098
0.1500   0.5784   0.2716   0.0355   0.0570   0.9075
0.1600   0.5677   0.2723   0.0380   0.0568   0.9052
0.1700   0.5569   0.2731   0.0405   0.0567   0.9028
0.1800   0.5461   0.2739   0.0431   0.0565   0.9005
0.1900   0.5352   0.2748   0.0456   0.0563   0.8981
0.2000   0.5243   0.2757   0.0482   0.0561   0.8957
0.2100   0.5133   0.2767   0.0509   0.0558   0.8933
0.2200   0.5023   0.2777   0.0536   0.0556   0.8909
0.2300   0.4911   0.2789   0.0563   0.0553   0.8884
0.2400   0.4799   0.2801   0.0591   0.0550   0.8859
0.2500   0.4686   0.2814   0.0619   0.0547   0.8833
0.2600   0.4573   0.2827   0.0648   0.0544   0.8807
0.2700   0.4458   0.2842   0.0678   0.0541   0.8781
0.2800   0.4342   0.2858   0.0708   0.0538   0.8754
0.2900   0.4225   0.2875   0.0739   0.0534   0.8727
0.3000   0.4107   0.2893   0.0771   0.0530   0.8698
0.3100   0.3987   0.2913   0.0804   0.0526   0.8670
0.3200   0.3866   0.2934   0.0838   0.0522   0.8640
0.3300   0.3743   0.2957   0.0874   0.0517   0.8609
0.3400   0.3619   0.2981   0.0910   0.0512   0.8578
0.3500   0.3492   0.3008   0.0948   0.0507   0.8545
0.3600   0.3363   0.3037   0.0988   0.0501   0.8511
0.3700   0.3232   0.3068   0.1029   0.0495   0.8475
0.3800   0.3098   0.3102   0.1073   0.0489   0.8438
0.3900   0.2960   0.3140   0.1119   0.0482   0.8399
0.4000   0.2818   0.3182   0.1169   0.0474   0.8357
0.4100   0.2672   0.3228   0.1222   0.0466   0.8312
0.4200   0.2520   0.3280   0.1279   0.0457   0.8264
0.4300   0.2360   0.3340   0.1342   0.0447   0.8211
0.4400   0.2191   0.3409   0.1413   0.0435   0.8152
0.4500   0.2009   0.3491   0.1493   0.0422   0.8085
0.4600   0.1808   0.3592   0.1590   0.0405   0.8005
0.4700   0.1571   0.3729   0.1715   0.0384   0.7901
0.4800   0.1236   0.3964   0.1922   0.0349   0.7729
0.4810   0.1184   0.4006   0.1958   0.0342   0.7699
0.4820   0.1118   0.4062   0.2006   0.0334   0.7659
0.4830   0.0987   0.4183   0.2109   0.0317   0.7574
0.4831   NO SOLUTION

Fig. 10. Output from program TERNGAP

e) Listing of Program TERNGAP

Frank Turpin, S. Saxena

```
   1          DIMENSION CODE(3)
   2       10 READ(5,300,END=140) CODE,T,E12,E21,E13,E31,E23,E32,C
   3          WRITE(6,310) CODE,T,E12,E21,E13,E31,E23,E32,C
   4          STEP=0.01
   5          CALL BIGAP(T,E12,E21,Z10,Z20,KON)
   6          IF(KON)20,20,30
   7       20 WRITE(6,320) Z10,Z20
   8          GO TO 40
   9       30 WRITE(6,330)
  10          GO TO 10
  11       40 CALL BIGAP(T,E13,E31,U1,U2,KON)
  12          IF(KON) 50,50,60
  13       50 WRITE(6,340) U1,U2
  14       60 CALL BIGAP (T,E23,E32,V1,V2,KON)
  15          IF(KON) 70,70,80
  16       70 WRITE(6,350)V1,V2
  17       80 X1=0.0
  18          X20=STEP
  19          WRITE(6,360)
  20          Y1=1.0-X1-Z10
  21          Y2=1.0-X1-Z20
  22          WRITE(6,370) X1,Y1,Z10,X1,Y2,Z20
  23          KON=0
  24      100 X1=X1+STEP
  25          CALL BINOD(T,E12,E21,E13,E31,E23,E32,C,Z10,X20,Z20,X1,Y1,Z1,
  25         $X2,Y2,Z2,KON,NG)
  26      104 IF(NG) 105,120,130
  27      105 IF(KON) 110,120,115
  28      115 X1=X1-STEP
  29          STEP=0.001
  30          GO TO 100
  31      110 X1=X1-STEP
  32          STEP=0.0001
  33          GO TO 100
  34      120 WRITE(6,370)X1,Y1,Z1,X2,Y2,Z2
  35          GO TO 100
  36      130 WRITE(6,380) X1
  37          GO TO 10
  38      300 FORMAT(3A3,8F8.0)
  39      310 FORMAT(////'1EXITED GAP - ',3A3,3X,'T=',F8.0,2X,'E12=',F8.0,
             $2X,'E21=',F8.0/2X,'E13=',F8.0,2X,'E31=',F8.0,2X,'E23=',F8.0,2X,
             $'E32=',F8.0,2X,'C=',F8.0)
  40      320 FORMAT(//' EDGE 1 SOLUTION - ',2F8.3)
  41      330 FORMAT(//' NO SOLUTION - WRONG STARTING EDGE')
  42      340 FORMAT(//' EDGE 2 SOLUTION - ',2F8.3)
  43      350 FORMAT(//' EDGE 3 SOLUTION - ',2F8.3)
  44      360 FORMAT(//,4X,'TIE LINE COORDINATES',//4X,'X1',5X,'Y1',5X,'Z1',
             $5X,'X2',5X,'Y2',5X,'Z2'/)
  45      370 FORMAT(6F8.4)
  46      380 FORMAT(F8.4,2X,'NO SOLUTION')
  47      140 STOP
  48          END

  49          SUBROUTINE BINOD(T,E13,E31,E12,E21,E23,E32,C,Z10,X20,Z20,
             2X1,Y1,Z1,X2,Y2,Z2,KON,NG)
  50          NG=0
  51          ITER=0
  52          RT=1.987*T
  53          A21=E21-E12
  54          A31=E31-E13
  55          A23=E23-E32
  56          A2131=E21+E12+E31+E13-E23-E32
  57          A3221=E32+E23+E21+E12-E31-E13
  58          A3132=E31+E13+E32+E23-E21-E12
  59          AA1=A21+A31
  60          AA2=A23+A21
  61          AA3=A23-A31
  62          X2=X20
  63          Z1=Z10
```

```
64              Z2=Z20
65              Y1=1-X1-Z1
66              IF(Y1.LT.0.001) GO TO 13
67              Y2=1-X2-Z2
68         1    Z1=1.0-X1-Y1
69              Z2=1.0-X2-Y2
70              XX=X2**2-X1**2
71              YY=Y2**2-Y1**2
72              ZZ=Z2**2-Z1**2
73              XXZ=X2**2*Z2-X1**2*Z1
74              XXY=X2**2*Y2-X1**2*Y1
75              YYX=Y2**2*X2-Y1**2*X1
76              YYZ=Y2**2*Z2-Y1**2*Z1
77              ZZX=Z2**2*X2-Z1**2*X1
78              ZZY=Z2**2*Y2-Z1**2*Y1
79              XY=X2*Y2-X1*Y1
80              YZ=Y2*Z2-Y1*Z1
81              ZX=Z2*X2-Z1*X1
82              XY2=X1*Y1+X2*Y2
83              YZ2=Y1*Z1+Y2*Z2
84              ZX2=Z1*X1+Z2*X2
85              XYZ=X2*Y2*Z2-X1*Y1*Z1
86              XYZ2=X2*Y2*Z2+X1*Y1*Z1
87              GXO=RT*ALOG(X2/X1)
88              GYO=RT*ALOG(Y2/Y1)
89              GZO=RT*ALOG(Z2/Z1)
90              GI=GZO+E12*XX+2.0*A21*XXZ+E13*YY+2.0*A31*YYZ+0.5*A2131*XY
                2+(A21+A31)*XYZ+A23*(XXY-YYX)-C*(XY2-2.0*XYZ2)
91              GJ=GXO+E23*YY-2.0*A23*YYX+E21*ZZ-2.0*A21*ZZX+0.5*A3221*YZ
                2-(A23+A21)*XYZ+A31*(YYZ-ZZY)-C*(YZ2-2.0*XYZ2)
92              GK=GYO+E31*ZZ-2.0*A31*ZZY+E32*XX+2.0*A23*XXY+0.5*A3132*ZX+
                2(A23-A31)*XYZ-A21*(ZZX-XXZ)-C*(ZX2-2.0*XYZ2)
93              DGIDX2=-RT/Z2-6.C*A21*X2**2+2.0*(E12+2.0*A21)*X2+2.0*(-2.0*A21-
                2AA1+A23-2.0*C)*X2*Y2+(0.5*A2131+AA1+C)*Y2+(-2.0*A31-AA1-A23-2.0
                3*C)*Y2**2
94              DGIDY1=RT/Z1+(2.0*A21+AA1-A23-2.0*C)*X1**2+(-0.5*A2131-AA1+C)
                2*X1+2.0*(2.0*A31+AA1+A23-2.0*C)*X1*Y1+2.0*(-E13-2.0*A31)*Y1+
                36.0*A31*Y1**2
95              DGIDY2=-RT/Z2+(-2.0*A21-AA1+A23-2.0*C)*X2**2+(0.5*A2131+AA1+C)
                2*X2+2.*(-2.0*A31-AA1-A23-2.0*C)*X2*Y2+2.0*(E13+2.0*A31)*Y2
                3-6.0*A31*Y2**2
96              DGJDX2=RT/X2-6.0*A21*X2**2+2.0*(E21+4.0*A21)*X2+(-2.*A21-2.0*E21)
                2+2.0*(-3.0*A21+A23-A31-2.0*C)*X2*Y2+(2.0*E21+3.0*A21-0.5*A3221
                3-A23+2.0*A31+3.0*C)*Y2+(-A23-A21-3.0*A31-2.0*C)*Y2**2
97              DGJDY1=(3.0*A21-A23+A31-2.0*C)*X1**2+(-2.0*E21-3.0*A21+0.5*A3221
                2+A23-2.*A31+3.*C)*X1+2.*(A23+A21+3.*A31-2.0*C)*X1*Y1
                3+(2.0*E21-0.5*A3221+A31-C)*Y2+2.0*(-E23-E21+0.5*A3221-3.0*A31+C)
                4*Y1+3.0*(2.0*A31)*Y1**2
98              DGJDY2=(-3.0*A21+A23-A31-2.0*C)*X2**2+(2.0*E21+3.0*A21-0.5*A3221
                2-A23+2.0*A31+3.0*C)*X2+2.0*(-A23-A21-3.0*A31-2.0*C)*X2*Y2+
                3(-2.0*E21+0.5*A3221-A31-C)+2.0*(E23+E21-0.5*A3221+A31+C)*Y2
99              DGKDX2=-6.*A21*X2**2+2.0*(E31+E32-0.5*A3132+3.0*A21+C)*X2+
                2(0.5*A3132-A21-C-2.0*E31)+2.0*(-A31+A23+3.0*A21-2.0*C)*X2*Y2
                3+(5.0*A31-0.5*A3132+A23+3.0*C+2.0*E31)*Y2+(-3.0*A31-A23-A21
                4-2.0*C)*Y2**2
100             DGKDY1=-RT/Y1+(A31-A23+3.0*A21-2.0*C)*X1**2+(-2.0*E31-3.0*A31
                2+0.5*A3132-A23-2.0*A21+3.0*C)*X1+2.0*(3.0*A31+A23+A21-2.0*C)
                3*X1*Y1+(2.0*E31+2.0*A31)+2.0*(-E31-4.0*A31)*Y1+6.0*A31*Y1**2
101             DGKDY2=RT/Y2+(-A31+A23+3.0*A21-2.0*C)*X2**2+(5.0*A31-0.5*A3132
                2+A23+3.0*C+2.0*E31)*X2+2.0*(-3.0*A31-A23-A21-2.0*C)*X2*Y2
                3+(2.0*E31-2.0*A31)+2.0*(E31+4.0*A31)*Y2-6.0*A31*Y2**2
102             DEL=DGIDX2*DGJDY1*DGKDY2+DGIDY1*DGJDY2*DGKDX2+DGIDY2*DGJDX2*DGKDY1
                2-DGIDY2*DGJDY1*DGKDX2-DGIDX2*DGJDY2*DGKDY1-DGIDY1*DGJDX2*DGKDY2
103             DX2=(-GI*DGJDY1*DGKDY2-DGIDY1*DGJDY2*GK-DGIDY2*GJ*DGKDY1+GK*DGJDY1
                2*DGIDY2+DGKDY1*DGJDY2*GI+DGKDY2*GJ*DGIDY1)/DEL
104             DY1=(-DGIDX2*GJ*DGKDY2-GI*DGJDY2*DGKDX2-DGIDY2*DGJDX2*GK+
                2DGKDX2*GJ*DGIDY2+GK*DGJDY2*DGIDX2+DGKDY2*DGJDX2*GI)/DEL
105             DY2=(-DGIDX2*DGJDY1*GK-DGIDY1*GJ*DGKDX2-GI*DGJDX2*DGKDY1+
                2DGKDX2*DGJDY1*GI+DGKDY1*GJ*DGIDX2+GK*DGJDX2*DGIDY1)/DEL
106             IF(DY1.GT.0.10) DY1=0.10
107             IF(DY2.GT.0.10) DY2=0.10
108             IF(DX2.GT.0.10) DX2=0.10
109             IF(ABS(DX2)-1.E-04*ABS(X2))2,2,4
110        2    IF(ABS(DY1)-1.E-04*ABS(Y1))3,3,4
111        3    IF(ABS(DY2)-1.E-04*ABS(Y2))6,6,4
```

```
112      4 X2=X2+DX2
113        Y1=Y1+DY1
114        Y2=Y2+DY2
115        IF(X2.LE.0.0) X2=0.001
116        IF(Y1.LE.0.0) Y1=0.001
117        IF(Y2.LE.0.0) Y2=0.001
118        ITER=ITER+1
119        IF(ITER-50)1,1,13
120      6 IF(Y1)13,13,7
121      7 IF(Z1)13,13,8
122      8 IF(X2)13,13,9
123      9 IF(Y2)13,13,10
124     10 IF(Z2)13,13,11
125     11 IF(X1)13,13,12
126     12 IF(ABS(X1-X2) .GT. 1.0E-03) GO TO 14
127        IF(ABS(Y1-Y2) .GT. 1.0E-03) GO TO 14
128        GO TO 13
129     14 Z10=Z1
130        X20=X2
131        Z20=Z2
132        RETURN
133     13 NG=-1
134        IF(KON.EQ.-1) NG=1
135        IF(KON.EQ. 1) KON=-1
136        IF(KON.EQ. 0) KON=1
137        RETURN
138        END


139        SUBROUTINE BIGAP(T,E12,E21,X1,X2,KON)
140        READ(5,300) X1,X2
141    300 FORMAT(2F8.4)
142        KON=0
143        RT=1.987*T
144        B=(E12+E21)/2.0
145        C=(E21-E12)/2.0
146        P=12.0*C/RT
147        Q=(18.0*C-2.0*B)/RT
148        R=(6.0*C-2.0*B)/RT
149        S=(2.0*B-6.0*C)/RT
150        ITER=0
151     30 F1=ALOG(X2/X1)+P/3.0*(X2**3-X1**3)+Q/2.0*(X1**2-X2**2)+R*(X2-X1)
152        F2=ALOG((1.0-X2)/(1.0-X1))+P/3.0*(X2**3-X1**3)+(B-3.0*C)/RT
         $*(X2**2-X1**2)
153        DF1DX1=-1.0/X1-P*X1**2+Q*X1-R
154        DF1DX2=1.0/X2+P*X2**2-Q*X2+R
155        DF2DX1=1.0/(1.0-X1)-P*X1**2-S*X1
156        DF2DX2=-1.0/(1.0-X2)+P*X2**2+S*X2
157        DELX1=(DF1DX2*F2-DF2DX2*F1)/(DF1DX1*DF2DX2-DF1DX2*DF2DX1)
158        DELX2=(DF2DX1*F1-DF1DX1*F2)/(DF1DX1*DF2DX2-DF1DX2*DF2DX1)
159        IF(ABS(DELX1)-1.0E-04*ABS(X1)) 20,20,50
160     20 IF(ABS(DELX2)-1.0E-04*ABS(X2)) 40,40,50
161     50 ITER=ITER+1
162        IF(ITER-50) 60,60,100
163     60 X1=X1+DELX1
164        X2=X2+DELX2
165        IF(X1.LE.0.0) X1=0.005
166        IF(X2.LE.0.0) X2=0.005
167        IF(X1.GE.1.0) X1=0.995
168        IF(X2.GE.1.0) X2=0.995
169        IF(ABS(X1-X2).LE.1.0E-03) GO TO 100
170        GO TO 30
171     40 RETURN
172    100 KON=1
173        RETURN
174        END
```

References

AHRENS, L. H.: The use of ionization potentials. Part I. Ionic radii of the elements. Geochim. Cosmochim. Acta **2**, 155 (1952).

AKIMOTO, S., KATURA, T., SYONO, Y., FUJISAWA, H., KOMADA, E.: Polymorphic transitions of pyroxenes $FeSiO_3$ and $CoSiO_3$ at high pressures and temperatures. J. Geophys. Res. **70**, 5269—5278 (1965).

ALBEE, A. L.: Phase equilibria in three assemblages of kyanite-zone pelitic schists, Lincoln mountain quadrangle, central vermont. J. Petrol. **6**, 246—301 (1965).

ALLMANN, R., HELLNER, E.: Indirect calculation of the partial free energy of albite and orthoclase in alkali feldspar. Norsk Geol. Tidsskr. **42**, 346—348 (1962).

ANNERSTEN, H.: A mineral chemical study of a metamorphosed iron formation in Northern Sweden. Lithos **1**, 374—397 (1968).

ANNERSTEN, H., EKSTRÖM, T.: Distribution of major and minor elements in coexisting minerals from a metamorphosed iron formation. Lithos **4**, 185—204 (1971).

BACHINSKI, SHARON, W., MÜLLER, G.: Experimental determinations of the microcline-low albite solvus. J. Petrol. **12**, 329—356 (1971).

BANNO, S.: Classification of eclogites in terms of physical conditions of their origin. Phys. Earth Planet. Interiors **3**, 405—421 (1970).

BARTH, T. F. W.: Studies in gneiss and granite I and II. Nor. Vid-Akad. Skr. No. 1 (1956).

BARTH, T. F. W.: Feldspar solid solutions. Chem. Erde **22**, 14—20 (1962).

BETHKE, P. M., BARTON, P. B. JR.: Distribution of minor elements between coexisting sulphide minerals. Econ. Geol. **66**, 140—163 (1971).

BINNS, R. A.: Metamorphic pyroxenes from the Broken Hill District, New South Wales. Mineral. Mag. **259**, 320—338 (1962).

BIRLE, J. D., GIBBS, G. V., MOORE, P. B., SMITH, J. V.: Crystal structures of natural olivines. Am. Mineral. **53**, 807—824 (1968).

BONNICHSEN, B.: Metamorphic pyroxenes and amphiboles in the Bivabik iron formation, Dunka River area, Minnesota. Mineral. Soc. Am. Spec. Pap. **2**, 217—239 (1969).

BORGHESE, C.: Cation distributions in multisublattice ionic crystals and applications to solid solutions of ferromagnetic garnets and spinels. J. Phys. Chem. Solids **28**, 2225—2227 (1967).

BOWEN, N. L., SCHAIRER, J. F.: The system $MgO-FeO-SiO_2$. Amer. J. Sci. 151—217 (1935).

BOWEN, N. L., TUTTLE, O. F.: The system $NaAlSi_3O_8-KAlSi_3O_8-H_2O$. J. Geol. **58**, 489 (1950).

BOYD, F. R., SCHAIRER, J. F.: The system $MgSiO_3-CaMgSi_2O_6$. J. Petrol. **5**, 275—309 (1964).

BRADLEY, R. S.: Thermodynamic calculations on phase equilibria involving fused salts. Pt. II. Solid solutions and application to olivines. Am. J. Sci. **260**, 550—554 (1962).

BROWN, W. L.: The crystallographic and petrologic significance of peristerite unmixing in the acid plagioclases. Z. Krist. **113**, 330—340 (1960).

BROWN, W. L.: Peristerite unmixing in the plagioclases and metamorphic facies series. Norsk Geol. Tidsskr. **42**, 354—382 (1962).

BROWN, W. L.: Entropie de configuration, solution solide et structure cristalline. Bull. Soc. Franc. Mineral. Crist. **94**, 38—44 (1971).

BROWN, E. H.: The green schist facies in part of eastern Otago, New Zealand. Contrib. Mineral. Petrol. **14**, 259—292 (1967).

BROWN, E. H.: Some zoned garnets from the green schist facies. Am. Mineral. **54**, 1662—1677 (1970).

BURNS, R. G., FYFE, W. S.: Distribution of elements in geological processes. Chem. Geol. **1**, 49—56 (1966).

BUTLER, P., JR.: Mineral compositions and equilibria in the metamorphosed iron formation of the Gagnon Region, Quebec, Canada. J. Petrol. **10**, 56—101 (1969).

CARLSON, H. C., COLBURN, A. P.: Vapour-liquid equilibria of non-ideal solutions. Ind. Eng. Chem. **34**, 581—589 (1942).

CHANG, L. L. Y.: Solid solutions of scheelite with other $R^{II}WO_4$-type tungstates. Am. Mineral. **52**, 427—435 (1967).

CLARK, J. R., APPLEMAN, D. E., PAPIKE, J. J.: Crystal-chemical characterization of clino-pyroxenes based on eight new structure refinements. Mineral. Soc. Spec. Pap. **2**, 31—50 (1969).

CRAWFORD, M. L.: Composition of plagioclase and associated minerals in some schists from Vermont, USA, and South Westland, New Zealand, with inferences about peristerite solvus. Contrib. Mineral. Petrol. **13**, 269—294 (1966).

DENBIGH, K.: The Principles of Chemical Equilibrium. Cambridge: Univ. Press 1966.

DENNEN, W. H., BLACKBURN, W. H., QUESADA, A.: Aluminium in quartz as a geothermometer. Contrib. Mineral. Petrol. **27**, 332—342 (1970).

DIENES, G. J.: Kinetics of order-disorder transformations. Acta Met. **3**, 549—557 (1955).

DIXON, W. T.: Biomedical Computer Programs. DIXON, W. J. (ed.): Berkeley: University of California Press 1970.

EKSTRÖM, T. K.: The distribution of fluorine among some coexisting minerals. Contr. Mineral. Petrol. **34**, 192—200 (1972).

EVANS, B. J., GHOSE, S., HAFNER, S.: Hyperfine splitting of ^{57}Fe and Mg-Fe order-disorder in orthopyroxenes ($MgSiO_3$-$FeSiO_3$ solid solution). J. Geol. **75**, 306—322 (1967).

FINGER, L. W.: Fe/Mg ordering in olivines. Carnegie Inst. Wash. Yearbook **69**, 302—304 (1970).

GENTILE, A. L., ROY, R.: Isomorphism and crystalline solubility in the garnet family: Am. Mineral. **45**, 701—711 (1960).

GHOSE, S.: The crystal structure of a cummingtonite. Acta Cryst. **14**, 622—627 (1961).

GHOSE, S.: The nature of Mg^{2+}-Fe^{2+} distribution in some ferromagnesian minerals. Am. Mineral. **47**, 388—394 (1962).

GHOSE, S.: $Mg^{2+}Fe^{2+}$ order in an orthopyroxene, $Mg_{0.93}Fe_{1.07}Si_2O_6$. Z. Krist. **122**, 81—99 (1965).

GOLDSMITH, J. R., NEWTON, R. C.: P-T-X relations in the system $CaCO_3$-$MgCO_3$ at high temperatures and pressures. Am. J. Sci. **267a**, 160—190 (1969).

GORBATSCHEV, R.: Element distribution between biotite and Ca-amphibole in some igneous or pseudo-igneous plutonic rocks. Neues Jahrb. Mineral. Abhandl. **111**, 314—342 (1969).

GREEN, E. J.: Predictive thermodynamic models for mineral systems. I. Quasi-chemical analysis of the halite-sylvite subsolidus. Am. Mineral. **55**, 1692—1713 (1970).

GREENWOOD, H. J.: The N-dimensional tie-line problems. Geochim. Cosmochim. Acta **31**, 465—490 (1967).

GROVER, J. E., ORVILLE, P. M.: The partitioning of cations between coexisting single- and multi-site phases with application to the assemblages: orthopyroxene-clinopyroxene and orthopyroxene-olivine. Geochim. Cosmochim. Acta **33**, 205—226 (1969).

GUGGENHEIM, E. A.: Theoretical basis of Raoult's law. Trans. Faraday Soc. **33**, 151—159 (1937).

GUGGENHEIM, E. A.: Mixtures. Oxford: Clarendon Press 1952.

GUGGENHEIM, E. A.: Thermodynamics. Amsterdam: North-Holland Publ. Co. 1967.

HAFNER, S. S., VIRGO, D., WARBURTON, D.: Cation distributions and cooling history of clinopyroxenes from Oceanus Procellarum. Proc. Second Lunar Sci. Conf. **1**, 91—108 (1971).

HIETANEN, A.: Distribution of elements in biotite-hornblende pairs and in an ortho-pyroxene-clinopyroxene pair from zoned plutons, northern Sierra Nevada, California. Contrib. Mineral. Petrol. **30**, 161—176 (1971).

HOVIS, G.: Thermodynamic properties of monoclinic potassium feldspars. Ph. D. Thesis. Cambridge, Massachusetts: Harvard University 1971.

HRICHOVA, R.: Contributions to the synthesis of garnets. I. Synthesis of spessartite and its analogues. Sci. Papers, Inst. Chem. Tech., Prague **G10**, 19—34 (1968).

INGERSON, E.: Geologic thermometry. Geol. Soc. Am. Spec. Pap. **62**, 465—488 (1955).

KAUFMANN, L., BERNSTEIN, H.: Computer Calculation of Phase Diagrams. New York-London: Academic Press, Inc. 1970.

KERN, R., WEISBROD, A.: Thermodynamics for Geologists. San Francisco: Freeman, Cooper & Co. 1967.

KING, M. B.: Phase Equilibrium in Mixtures. Oxford: Pergamon Press, Ltd. 1969.

KISCH, H., WARNAARS, F. W.: Distribution of Mg and Fe in cummingtonite-hornblende and cummingtonite-actinolite pairs from metamorphic assemblages. Contrib. Mineral. Petrol. **24**, 245—267 (1969).

KITAYAMA, K., KATSURA, T.: Activity measurements in orthosilicate and metasilicate solid solutions. I. Mg_2SiO_4-Fe_2SiO_4 and $MgSiO_3$-$FeSiO_3$ at $1204°$ C 1968.

KLEIN, C., (JR.): Coexisting amphiboles. J. Petrol. **9**, 281—330 (1968).

KRANCK, S. H.: A study of phase equilibria in a metamorphic iron formation. J. Petrol. **2**, 137—184 (1961).

KRETZ, R.: Chemical study of garnet, biotite and hornblende from gneisses of southwestern Quebec, with emphasis on distribution of elements in coexisting minerals. J. Geol. **67**, 371—402 (1959).

KRETZ, R.: Some applications of thermodynamics to coexisting minerals of variable composition. Examples: orthopyroxene-clinopyroxene and orthopyroxene-garnet. J. Geol. **69**, 361—387 (1961).

KRETZ, R.: Distribution of magnesium and iron between orthopyroxene and calcic pyroxene in natural mineral assemblages. J. Geol. **71**, 773—785 (1963).

KRETZ, R.: Analysis of equilibrium in garnet-biotite-sillimanite gneisses from Quebec. J. Petrol. **5**, 1—20 (1964).

KULLERUD, G.: The FeS-ZnS system as a geological thermometer. Norsk. Geol. Tidsskr. **32**, 61—147 (1953).

KUREPIN, V. A.: Conditions for stability of the $MgSiO_3$-$FeSiO_3$ pyroxenes. Geokhimiya **8**, 1000—1004 (1970).

182   References

LARIMER, J. W.: Experimental studies on the system $Fe-MgO-SiO_2-O_2$ and their
  bearing on petrology of chondritic meteorites. Geochim. Cosmochim. Acta **32**,
  1187—1207 (1968).
LAVES, F., GOLDSMITH, J. R.: The effect of temperature and composition on the Al-
  Si distribution in anorthite. Z. Krist. **106**, 227—235 (1955).
LEELANANDAM, C.: Chemical mineralogy of hornblendes and biotites from the
  charnockitic rocks of Kondapalli, India. J. Petrol. **11**, 475—505 (1970).
LINDSLEY, D. H.: Ferrosilite. Carnegie Inst. Wash. Yearbook **64**, 148 (1965).
LINDSLEY, D. H., MACGREGOR, I. D., DAVIS, B. T. C.: Synthesis and stability of ferro-
  silite. Carnegie Inst. Wash. Yearbook **63**, 174—176 (1964).
LOEWENSTEIN, W.: The distribution of aluminum in the tetrahedra of silicates and
  aluminates. Am. Mineral. **39**, 92—96 (1954).
LUPIS, C. H. P., ELLIOTT, J. F.: Generalized interaction coefficients. Part II: Free
  energy terms and the quasi-chemical theory. Acta Met. **14**, 1019—1032 (1966).
LUTH, W. C., TUTTLE, O. F.: The alkali feldspar solvus in the system Na_2O-K_2O-
  $Al_2O_3-SiO_2-H_2O$. Am. Mineral. **51**, 1359—1373 (1966).
MATSUI, Y., BANNO, S.: Intracrystalline exchange equilibrium in silicate solid solu-
  tions. Proc. Jap. Acad. **41**, 461—466 (1965).
MATSUMOTO, T.: Pressure-temperature conditions for the formation of peridotite
  inclusions — an application of a regular solution model to partitioning of Mg,
  Fe and Co between coexisting olivine and orthopyroxene. Geochem. J. **4**, 111—
  121 (1971).
MCCALLUM, I. S.: Equilibrium relationships among the coexisting minerals in the
  Stillwater, Complex, Montana. University of Chicago 1968.
MCINTYRE, W. L.: Trace element partition coefficients — a review of theory and
  applications to geology. Geochim. Cosmochim. Acta **27**, 1209—1264 (1963).
MEDARIS, L. G., (JR.): Partitioning of Fe^{2+} and Mg^{2+} between coexisting synthetic
  olivine and orthopyroxene. Am. J. Sci. **267**, 945—968 (1969).
MUELLER, R. F.: Compositional characteristics and equilibrium relations in mi-
  neral assemblages of a metamorphosed iron formation. Am. J. Sci. **258**, 449—
  493 (1960).
MUELLER, R. F.: Energetics of certain silicate solutions. Geochim. Cosmochim.
  Acta **26**, 581—598 (1962).
MUELLER, R. F.: Theory of immiscibility in mineral systems. Mineral. Mag. **33**,
  1015—1023 (1964).
MUELLER, R. F.: Stability relations of the pyroxenes and olivine in certain high
  grade metamorphic rocks. J. Petrol. **7**, 363—374 (1966).
MUELLER, R. F.: Model for order-disorder kinetics in certain quasi-binary crystals
  of continuously variable composition. J. Phys. Chem. Solids **28**, 2239—2243
  (1967).
MUELLER, R. F.: Kinetics and thermodynamics of intracrystalline distributions. Mineral. Soc. Am. Spec. Pap. **2**, 83—93 (1969).
MUELLER, R. F.: Two-step mechanism for order-disorder kinetics in silicates. Am.
  Mineral. **55**, 1210—1218 (1970).
MUELLER, R. F., GHOSE, S., SAXENA, S.: Partitioning of cations between coexisting
  single- and multi-site phases: A discussion. Geochim. Cosmochim. Acta **34**,
  1356—1360 (1970).
NAFZIGER, R. H., MUAN, A.: Equilibrium phase compositions and thermodynamic
  properties of olivines and pyroxenes in the system $MgO-FeO-SiO_2$. Am. Mineral. **52**, 1364—1385 (1967).
NAVROTSKY, A., KLEPPA, O. J.: The thermodynamics of cation distributions in simple spinels. J. Inorg. Nucl. Chem. **29**, 2701—2714 (1967).

OLSEN, E., BUNCH, T. E.: Empirical derivation of activity coefficients for the magnesium-rich portion of the olivine solid solution: Am. Mineral. **55**, 1829—1842 (1970).

OLSEN, E., MUELLER, R. F.: Stability of orthopyroxenes with respect to pressure, temperature and composition. J. Geol. **74**, 620-625 (1966).

ORVILLE, P. M.: Alkali ion exchange between vapor and feldspar phases. Am. J. Sci. **261**, 201—237 (1963).

ORVILLE, P. M.: Plagioclase cation exchange equilibria with aqueous chloride solution at 700° C and 2000 bars in the presence of quartz. Am. J. Sci. **272**, 234—272 (1972).

PERCHUK, L. L.: The paragenesis of nepheline with alkali feldspar as the indicator of mineral equilibrium thermodynamic conditions. Dokl. Akad. Nauk. SSSR **161**, No. 4 (1965).

PERCHUK, L. L., RYABCHIKOV, I. D.: Mineral equilibria in the system nepheline-alkali feldspar-plagioclase. J. Petrol. **9**, 123—167 (1968).

PERRY, K., (JR.): Construction of a single (m + 2) dimensional phase diagram from petrochemical data (in press, 1973).

PHILLIPS, M. W., RIBBE, P. H.: Bond length variations in monoclinic potassium-rich feldspars. Am. Mineral. (in press, 1973).

PRIGOGINE, I., DEFAY, R.: Chemical Thermodynamics. London: Longmans, Green & Co. Ltd. 1954.

RAHMAN, S., MACKENZIE, W. S.: The crystallization of ternary feldspars: A study from natural rocks. Am. J. Sci. **267 A**, 391—406 (1969).

RAMBERG, H.: The origin of metamorphic and metasomatic rocks. Univ. Chicago Press 1952a.

RAMBERG, H.: Chemical bonds and the distribution of cations in silicates. J. Geol. **60**, 331—355 (1952b).

RAMBERG, H.: Relative stabilities of some simple silicates as related to the polarization of the oxygen ions. Am. Mineral. **39**, 256—271 (1954).

RAMBERG, H.: Chemical thermodynamics in mineral studies. Vol. 5 of Physics and Chemistry of the Earth. Oxford: Pergamon Press, Inc. 1963.

RAMBERG, H., DEVORE, D. G. W.: The distribution of Fe^{2+} and Mg^{2+} in coexisting olivines and pyroxenes. J. Geol. **59**, 193—210 (1951).

RIBBE, P. H.: An X-ray and optical investigation of the peristerite plagioclases. Am. Mineral. **45**, 626—635 (1960).

RIBBE, P. H.: Observation on the nature of unmixing in peristerite plagioclases. Norsk. Geol. Tidsskr. **42**, 138—151 (1962).

RIBBE, P. H., MEGAW, H. D.: The structure of transitional anorthite — a comparison with primitive anorthite. Norsk. Geol. Tidsskr. **42**, 158—167 (1962).

RIBBE, P. H., GIBBS, G. V.: Statistical analysis and discussion of mean Al/Si-O bond distances and the aluminum content of tetrahedra in feldspars. Am. Mineral. **54**, 85—94 (1969).

RIBBE, P. H., STEWART, D. B., PHILLIPS, M. W.: Structural explanations for variations in the lattice parameters of sodic plagioclase. Geol. Soc. Am. Abs. **2**, 663 (1970).

RIBBE, P. H.: One parameter characterization of the average Al/Si distribution in plagioclase feldspars. J. Geophys. Res. (in press, 1972).

RUMBLE, D., III.: Thermodynamic analysis of phase equilibria in the system Fe_2TiO_4-Fe_3O_4-TiO_2. Carnegie Inst. Wash. Yearbook **69**, 198—206 (1970).

SAHAMA, TH. G., TORGESON, D. R.: Some examples of the application of thermochemistry to petrology. J. Geol. **57**, 255—262 (1949).

SAXENA, S. K.: Crystal-chemical aspects of distribution of elements among certain coexisting rock-forming silicates. Neues Jahrb. Mineral. Abhandl. **108**, 292—323 (1968a).

SAXENA, S. K.: Chemical study of phase equilibria in charnockites, Varberg, Sweden. Am. Mineral. **53**, 1674—1695 (1968 b).

SAXENA, S. K.: Distribution of elements between coexisting minerals and the nature of solid solution in garnet. Am. Mineral. **53**, 994—1014 (1968 c).

SAXENA, S. K.: Silicate solid solutions and geothermometry. 2. Distribution of Fe^{2+} and Mg^{2+} between coexisting olivine and pyroxene. Contrib. Mineral. Petrol. **22**, 147—156 (1969 a).

SAXENA, S. K.: Silicate solid solutions and geothermometry. 3. Distribution of Fe and Mg between coexisting garnet and biotite. Contrib. Mineral. Petrol. **22**, 259—267 (1969 b).

SAXENA, S. K.: Mg^{2+}-Fe^{2+} order-disorder in orthopyroxene and the Mg^{2+}-Fe^{2+} distribution between coexisting minerals. Lithos **4**, 345—356 (1971).

SAXENA, S. K., GHOSE, S.: Mg^{2+}-Fe^{2+} order-disorder and the thermodynamics of the orthopyroxene-crystalline solution. Am. Mineral. **56**, 532—559 (1971).

SAXENA, S. K., RIBBE, P. H.: Activity-composition relations in feldspars. Contr. Mineral. and Petrol. **37**, 131—138 (1972).

SCATCHARD, G., HAMER, W.: The application of equations for the chemical potentials to partially miscible solutions. J. Am. Chem. Soc. **57**, 1805—1809 (1935).

SCHULIEN, S., FRIEDRISCHSEN, H., HELLNER, E.: Das Mischkristallverhalten des Olivins zwischen 450° und 650° C bei 1 kb Druck. Neues Jahrb. Mineral. Monatsh. **4**, 141—147 (1970).

SECK, H. A.: Koexistierende Alkalifeldspate und Plagioklase in System $NaAlSi_3O_8$-$KAlSi_3O_8$-$CaAl_2Si_2O_8$-H_2O bei Temperaturen von 650° C bis 900° C. Neues Jahrb. Mineral. Abhandl. **115**, 315—345 (1971 a).

SECK, H. A.: Der Einfluß des Drucks auf die Zusammensetzung koexistierender Alkalifeldspate und Plagioklase. Contrib. Mineral. Petrol. **31**, 67—86 (1971 b).

SEN, S. K., CHAKRABORTY, K. R.: Magnesium-iron exchange equilibrium in garnet-biotite and metamorphic grade. Neues Jahrb. Mineral. Abh. **108**, 181—207 (1968).

SMITH, F. G.: Physical Geochemistry, Addison-Wesley, 1963.

SMITH, J. V.: Physical properties of order-disorder structures with especial reference to feldspar minerals. Lithos **3**, 145—160 (1970).

SMITH, J. V., BAILEY, S. W.: Second review of Al-O and Si-O tetrahedral distances. Acta Cryst. **16**, 801—811 (1963).

STEWART, D. B., RIBBE, P. H.: Structural explanation for variations in cell parameters of alkali feldspar with Al/Si ordering. Am. J. Sci. **267 A**, 444-462 (1969).

STORMER, J. C., CARMICHAEL, I. S. E.: Fluorine-hydroxyl exchange in apatite and biotite: a potential igneous geothermometer. Contrib. Mineral. Petrol. **31**, 121—131 (1971).

THOMPSON, J. B., (JR.): Thermodynamic properties of simple solutions. In: Abelson, P. H., (Ed.): Vol. II of Researches in Geochemistry, pp. 340—361. New York: John Wiley & Sons, Inc. 1967.

THOMPSON, J. B., (JR.): Chemical reactions in crystals. Am. Mineral. **54**, 341—375 (1969).

THOMPSON, J. B., JR., WALDBAUM, D. R.: Mixing properties of sanidine crystalline solutions. I. Calculations based on ion-exchange data. Am. Mineral. **53**, 1965—1999 (1968).

THOMPSON, J. B., JR., WALDBAUM, D. R.: Analysis of the two-phase region halite-sylvite in the system NaCl-KCl. Geochim. Cosmochim. Acta **33**, 671—690 (1969 a).

THOMPSON, J. B., JR., WALDBAUM, D. R.: Mixing properties of sanidine crystalline solutions. III. Calculations based on two-phase data. Am. Mineral. **54**, 811—838 (1969 b).

TOLMAN, R. C.: The Principles of Statistical Mechanics, p. 599. London, Oxford: University Press 1959.

VAN NESS, H. C.: Classical Thermodynamics of Non-electrolytic Solutions. London: McMillan & Co. 1964.

VIRGO, D., HAFNER, S. S.: Fe^{2+}, Mg order-disorder in heated orthopyroxenes. Mineral. Soc. Am. Spec. Pap. **2**, 67—81 (1969).

WAGNER, R. C.: Thermodynamics of Alloys. Reading: Addison-Wesley Publ. Co. 1952.

WALDBAUM, D. R.: High-temperature thermodynamic properties of alkali feldspars. Contrib. Mineral. Petrol. **17**, 71—77 (1968).

WALDBAUM, D. R., THOMPSON, J. B., JR.: Mixing properties of sanidine crystalline solutions: IV. Phase diagrams from equations of state. Am. Mineral. **54**, 1274—1298 (1969).

WARNER, R. D.: Experimental investigations in the system CaO-MgO-SiO_2-H_2O. Ph. D. dissertation, Stanford University 1971.

WHITTAKER, E. J. W.: Factors affecting element ratio in the crystallization of minerals. Geochim. Cosmochim. Acta **31**, 2275—2288 (1967).

WILLIAMS, R. J.: Reaction constants in the system Fe-Mgo-SiO_2-O_2 at/atm between 900° and 1300° C: Experimental results. Am. J. Sci. **270**, 334—360 (1971).

YAROSHEVSKIY, A. A.: Structural ordering of isomorphous mixtures. Geokhimiya **8**, 945—956 (1970).

Additional References

BARRON, L. M.: Nonideal thermodynamic properties of H_2O—CO_2 mixtures for O4—2 Kb and 400—700° C. Contrib. Mineral. Petrol. **39**, 184 (1973).

BESWICK, A. E.: An experimental study of alkali metal distributions in feldspars and micas. Geochim. Cosmochim. Acta **37**, 183—208 (1973).

FISHER, J. R., ZEN, E-AN: Thermochemical calculations from hydrothermal phase equilibrium data and the free energy of H_2O. Am. J. Sci. **270**, 297—314 (1971).

FLEMING, P. D.: Mg-Fe distribution between coexisting garnet and biotite, and the status of fibrolite. Geol. Mag. **109**, 477—482 (1973).

GORDON, T. M.: Determination of internally consistent thermodynamic data from phase equilibrium experiments. J. Geol. **81**, 199—208 (1973).

HENSEN, B. J., GREEN, D. H.: Experimental study of the stability of cordierite and garnet in pelitic compositions at high pressures and temperatures. III. Contrib. Mineral. Petrol. **38**, 151—166 (1973).

KRZANOWSKI, W. J., NEWMAN, A. C. D.: Computer simulation of cation distribution in the octahedral layers of micas. Min. Mag. **38**, 926—935 (1972).

MEL'NIK, YU. P.: Thermodynamic parameters of compressed gases and metamorphic reactions involving water and carbon dioxide. Geokhimiya, **6**, 654—662 (1972).

MUELLER, R. F.: System CaO—MgO—FeO—SiO_2—C—H_2—O_2: Some correlations from nature and experiment. Am. J. Sci. **273**, 152—170 (1973).

OHASHI, H., HARIYA, YU.: Order-disorder of ferric iron and aluminum in Ca-rich clinopyroxene. Proc. Jap. Acy. **46**, 684—687 (1970).

PERCHUK, L. L.: The staurolite-garnet thermometer. Dokl. Akad. Nauk. SSSR. **186**, 1405—1407 (1969).

TALANTSEV, A. S.: The plagioclase-muscorite geothermometer. Dokl. Akad. Nauk. SSSR **196**, 1190—1195 (1970).

WALDBAUM, D. R.: The configurational entropies of $Ca_2MgSi_2O_7$—$Ca_2SiAl_2O_7$ melilites and related minerals. Contrib. Mineral. Petrol. **39**, 35—54 (1973).

ZEN, E-AN: Thermochemical parameters of minerals from oxygen-buffered hydrothermal equilibrium data: Method, application to annite and almandine. Contrib. Mineral. Petrol. **39**, 65—80 (1973).

Subject Index

Minerals, Rocks and Inorganic Materials

G. M. BROWN and D. H. LINDSLEY: Synthesis and Stability of Pyroxenes

J. J. FAWCETT: Synthesis and Stability of Chlorite and Serpentine

D. H. LINDSLEY and S. HAGGERTY: Synthesis and Stability of Iron-Titanium Oxides

Subseries "Isotopes in Geology":

D. C. GREY: Radiocarbon in the Earth System

H. P. TAYLOR: Oxygen and Hydrogen Isotopes in Petrology

Subseries "Crystal Chemistry of Non-Metallic Materials"

R. E. NEWNHAM and R. ROY: Crystal Chemistry Principles

R. ROY and O. MULLER: The Major Binary Structural Families